CURIOUS DEVICES
AND
MIGHTY MACHINES

CURIOUS DEVICES

AND

MIGHTY MACHINES

EXPLORING SCIENCE MUSEUMS

SAMUEL J.M.M. ALBERTI

REAKTION BOOKS

Published by Reaktion Books Ltd
Unit 32, Waterside
44–48 Wharf Road
London N1 7UX, UK

www.reaktionbooks.co.uk

First published 2022
Copyright © Samuel J.M.M. Alberti 2022

Printed and bound in Great Britain by TJ Books Ltd, Padstow, Cornwall

A catalogue record for this book is available from the British Library

ISBN 978 1 78914 639 4

CONTENTS

1 Not 'a copper robot from the golden age of sci-fi' but a radio frequency accelerating cavity from CERN's Large Electron–Positron Collider, operational from 1989 to 1995, donated by CERN to the science collection of National Museums Scotland.

INTRODUCTION

Consider the 2-metre, 2-tonne metal machine now housed in the National Museum of Scotland (illus. 1). In 2014 a journalist compared such a device to 'the bust of a copper robot from the golden age of sci-fi, with a bulbous round head, ribbed skin, red cyclopean eye and silver claws which project, raptor-style, from what would be its breast'.[1] And yet it is not science fiction but science fact. This apparently crude apparatus is actually just one part of a highly precise scientific instrument, and if we were to look inside, we would see some of the most highly machined copper in the world (illus. 2). It was one of 128 accelerating cavities that were once positioned around the vast loop of the Large Electron–Positron Collider at the largest laboratory in the world, the European Organization for Nuclear Research known as CERN in Geneva. Their job was to whip subatomic particles around the 27-kilometre (17 mi.) loop so that they smashed together, annihilating each other and forming new particles from their energy. The cavities served their time between 1989 and 1995, and their successors were replaced by the Large Hadron Collider, which CERN scientists famously used to discern the presence of the Higgs boson, the so-called 'God particle'.

CERN distributed elements of the retired Electron–Positron Collider around the world: not only to laboratories and universities, but to museums. This one is in Edinburgh, on display on the third floor of the museum in which I work. This may seem a curious

2 The view inside the Large Electron–Positron Collider cavity: precisely machined copper.

place for the collider. Are not museums about fancy goods or relics: ancient pots, dinosaurs, Renaissance masterworks? Surely they are repositories for the dead and gone, or the impossibly elegant, rather than for a chunk of recently obsolete engineering?

This book is about why this, and thousands of other science and engineering objects, can be found in museums: how we collect such things, and how they are used. As well as the curious copper cavity, we will encounter giant steam engines, tiny test tubes, delicate measuring devices, chunks of oil rigs, medieval navigation devices and more; as well as Edinburgh we will visit collections in London, Moscow, Chicago, Munich, Ottawa, Paris, Oslo and Washington, DC. This introductory chapter invites you into these museums: to explore their contents and to introduce you to their staff and their visitors. Lots of things go on in science museums and there are many different groups of people involved: in *Curious Devices and Mighty Machines*, curators will mostly be our guides through the collections.

What Is a Science Museum?

Each chapter of this book opens up particular kinds of spaces and practices in science museums. You may know some of the museums already, whether as a visitor, a student or a professional; but it is in unpacking the obvious that we come to understand important things. Furthermore, it will become clear as we look around that science museums are not actually science museums. Rather, for the most part, they are about industry and technology. Some of them own up to this in their title, like the bustling Museum of Science and Industry in Chicago; others do not, like the Science Museum in London. As one veteran curator observed,

> The number of great national museums that specialize in science is extremely small (arguably, there are only a handful in Western Europe and none at all in North America); and of these, the greater number ... were established as

3 Vehicles stored in part of one area of the vast collections facility of the Polytechnic Museum in Tekstilshchiki District, Moscow.

much from a concern for technology and industry as from any particular love of science per se.[2]

This curious discrepancy is partly because science is about principles, which are hard to collect and even harder to display, and also because of the very particular origins of these organizations. The giants in this field began with the intention of enhancing industrial education, including in Paris the Musée des arts et métiers (Museum of Arts and Crafts (or Trades)) and in Munich the Deutsches Museum von Meisterwerken der Naturwissenschaft und Technik (German Museum for Masterpieces of Science and Technology), known simply as the Deutsches Museum.

'Big and oily' technology therefore dominates many of these museums. Industry is not only useful, but concrete and visually striking, so will loom large in the chapters that follow. We do, of course, find some science in science museums, typically manifested as the 'brass and glass' of antique instrumentation. This is especially the case in overtly historical museums such as the

4 Blue Wing of the lively, multidisciplinary Boston Museum of Science.

venerable Museum of the History of Science in Oxford, or the Museo della Scienza e della Tecnologia Leonardo da Vinci in its Milanese monastery, the largest collection of its kind in Italy.[3] Transport is another common feature, again offering charismatic, solid things to display (illus. 3): we will therefore encounter vehicles en route. Science museums like the Deutsches Museum also often include agriculture, so there will be the occasional whiff of the farm; and many of these collections include health and well-being, so we will also swallow the occasional dose of medical museology.[4]

I therefore use the term 'science collection' throughout these pages as an umbrella for science, technology, transport, sometimes agriculture, a bit of medicine and, very occasionally, mathematics – a shorthand my curatorial colleagues consider a cardinal sin. In Oslo, the innovative and occasionally frightening Norsk Teknisk Museum (Norwegian Museum of Science and

Technology) is an excellent example of the span of one of these institutions: 'transport, aviation, the history of wood and metal industries, plastics in a modern society, clocks and watches, calculating machines and computers, as well as the history of energy, electricity, oil and gas [and also] the Norwegian Telecom Museum'.[5]

Not only does each museum have its own characteristic internal configuration; but also, to complicate matters, the edges of science museums are blurry. Most of them are hybrids, blending one kind of museum with another, with particular orientations depending on their geography and history. Consider the other kind of science museum, which probably deserves the name more: natural history collections housing the remains of animals, plants and rocks.[6] Their creepy-crawlies, their fur-and-feather collections, are quite different from big-and-oily or brass-and-glass; their millions upon millions of specimens comprise great archives of biodiversity, and many of them are sites for cutting-edge science. In some settings the two species combine, or at least sit side by side. The buzzing Boston Museum of Science (illus. 4), for example, has its origins in a nineteenth-century natural history society, and proclaims itself to be the first museum to have all the sciences under one roof: it boasts zoology, geology, live animals, technology, physical sciences, an IMAX cinema and a planetarium. My own organization, National Museums Scotland, is an example of a 'world' museum: that is, we have not only science and natural history, but art, archaeology and social history. Elsewhere, science sits within, or alongside, other elements of history, especially at the National Museum of American History (originally the Museum of History and Technology) in the Smithsonian Institution.

The Boston Museum of Science also includes a feature that will run through this volume: a science centre. Science museums and science centres, and their respective audiences, have sometimes been contrasted with each other: the dusty historical museum with its ageing showcases and visitors in contradistinction to the noisy science centre with children running amok among interactives devoted to current science (illus. 5). But in practice, not only do they have more in common than one might think, they also often coexist within the same institution. Both deal with science past and present; science centres have elements of collections within them (illus. 6); and most large science museums will have an interactive within or woven through their galleries, as for example the 'Wonderlabs' in the uk's national Science Museum Group. Science centres and science museums rub shoulders in the science engagement ecosystem; the European network Ecsite (originally the European Collaborative for Science, Industry and Technology Exhibitions) includes science centres and museums side by side.

Given these blurry edges between science museums, natural history collections and science centres, it is rather difficult to count them. Among the 30,000 or so museums around the world it is likely that more than 2,000 of them have significant science collections of the kind we are talking about here.[7] These numbers are growing, too, especially in China, where science is a source

5 Hands-on experience for young people at the water play area of KidSpark, for children up to the age of eight, in the Ontario Science Centre, Toronto.

6 An object in a science centre: the i-Limb bionic hand at the Glasgow Science Centre. Science museums and science centres have much in common.

of particular pride and power, and has been the subject of invest-ment this century.[8] Needless to say, not every one of the several thousand will feature in these pages, and neither will we travel to all corners of the world. Rather, the main players on this stage are a group of organizations holding these kinds of collections in Western Europe and North America. They vary according to geog-raphy and governance: some are run by nation states (like the Musée des arts et métiers); some by local authorities; some at arm's length by trustee boards (like the Science Museum Group); some by voluntary organizations; and many by universities (such as the MIT Museum in Cambridge, Massachusetts, illus. 7). Some are tiny, with a handful of staff; some, like the Deutsches Museum, have several branches as well as their famous flagship sites.

Despite this variety, the institutions we will meet on our journey have markedly similar aims and objectives.[9] In trying to under-stand what a science museum is, we need to consider not only what is in them, but what they do and who they are for. According

15

7 A university science collection: the Maria Tallchief robot (named after the dancer Maria Tallchief) in the 'Robots and Beyond' exhibition (2017) at the MIT Museum.

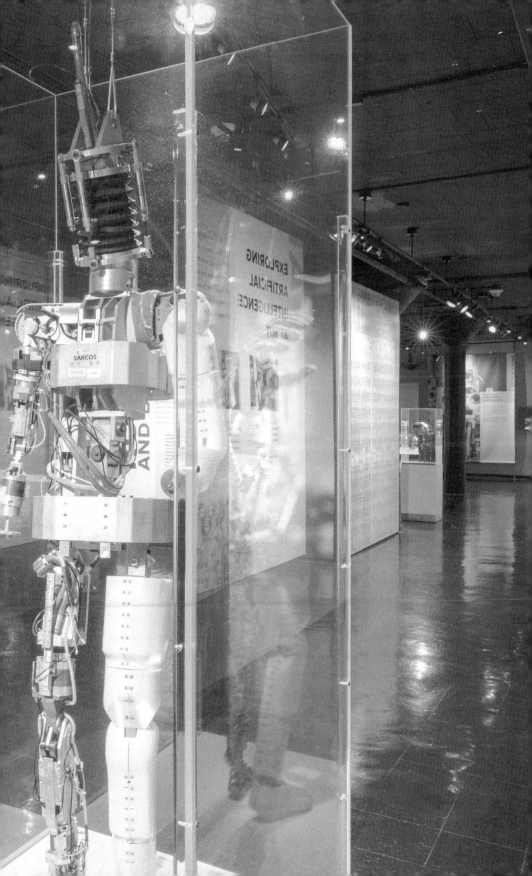

to the international museum organization ICOM, a museum is a 'non-profit, permanent institution in the service of society and its development, open to the public, which acquires, conserves, researches, communicates and exhibits the tangible and intangible heritage of humanity and its environment for the purposes of education, study and enjoyment'.[10] Science museums not only fulfil these criteria, but have specific elements in common.

First and foremost is their shared attention to their audiences; and especially, that they seek to *inspire* them – a term that comes up time and again. At National Museums Scotland, for instance, we 'create inspiring experiences that help our audiences' understanding of themselves and the world around them'.[11] We are a mixed museum, but for science collections generally these audiences are unusually young and the prevailing notion that science museums are exclusively for children will surface again and again. The Science Museum Group has an outstanding track record of attracting children to its sites – their London site is 'the number-one UK museum destination for school groups' – but more, their intention is that their 'offer and reputation for *lifelong* informal STEM learning and engagement will be the best in the world'.[12]

Science museums are also concerned with collecting and caring for their particular kinds of collections. Milan's Museo della Scienza e della Tecnologia seeks to achieve an 'international level of excellence in the protection, conservation, and enhancement of technical–scientific collections' and the National Museum of American History in Washington, DC, likewise aims to 'expand, strengthen and share our collections'.[13] Like any museum, the challenge here is how to balance this duty of care with the active use of the objects, between providing a service for today's visitors and ensuring they last for tomorrow's. As two wise colleagues reminded us when reflecting on scientific heritage, 'in the midst of so many regulations, norms, calibrations, it is easy to fall into the "preservationist trap" and forget why we do what we do in the first place. Ultimately, we preserve scientific heritage for the public.'[14]

As well as current and future audiences, museums also grapple with how to balance past and present – and, perhaps impossibly, the future – in the displays and collections. The vision of Ingenium, the organization that runs Canada's national science museums, is 'to inspire Canadians to celebrate and engage with their scientific, technological and innovative past, present and future'.[15] The Polytechnic Museum in Moscow likewise seeks to 'reveal the past, the present and the future of science'.[16] Curators try hard to escape the considerable historical gravity of collections by acquiring new material from the bewildering profusion of contemporary science, technology and medicine.

Science museums are also especially concerned with the relationship between science and society. Take the Deutsches Museum, which wants to be 'an outstanding place for communicating scientific and technical knowledge and for a constructive dialogue between science and society'.[17] Whereas other museums are clearly part of culture, because science is often considered to be opposed to it, those who manage science collections work harder to 'reflect science as a facet of broader culture'.[18] The Norsk Teknisk Museum in Oslo, for example, seeks 'to demonstrate the implications of the historical progress in science and technology, both socially and culturally',[19] just as the Rijksmuseum Boerhaave (the national science collection of the Netherlands, in Leiden), 'want[s] our visitors to discover the vital importance of science for everyday life'.[20] This opportunity-cum-tension, celebrating science as distinct from society, and yet embedded within the (rest of) culture, will reappear throughout the book.

A science museum, then, is a blend of gallery and playground; part library and (arguably) part shopping mall.[21] It seeks to inspire visitors (present and future) about science (past, present and future) and society. It does this with science objects.

What Is a Science Object?

It will become apparent that my concern in this book is not actually with science museums per se, but rather with collections of scientific material culture within them. I am interested in things: the physicality at the heart of museums. In the words of another

8 National Museums Scotland conservator Stuart McDonald sizing up the flare tip from the Murchison Oil Platform.

object-obsessed museologist, we will be considering 'real *stuff* and its three-dimensionality, weight, texture, surface temperature, smell, taste and spatio-temporal presence'.[22] In particular, I am fascinated by what these material qualities afford, the sensory encounter with them, the use of things. What sort of things are these, however? Asking a science curator what they have in their collection will usually generate an intake of breath and a response including 'diverse' or 'varied' or both. But then again, this is the response you would get if you asked any sort of curator about their collection. Let us nevertheless have a poke around the science collection to see what we might find.

To illustrate the categories of objects found in these collections, allow me to introduce half a dozen of my favourite things from museums around the world that will feature in the following chapters. We have already met the copper cavity from CERN, a large piece of kit that came from an even more massive scientific instrument. Other more manageably sized things in the brass-and-glass brigade include classic telescopes and microscopes,

9 Foucault's pendula are common features in traditional 'brass-and-glass' scientific instrument collections (in this case, more brass than glass): when suspended they demonstrate Earth's rotation. This 1883 example is in the Science Museum's collection.

clocks and computers, and demonstration apparatus, like the famous pendulum designed by French physicist Léon Foucault to show the rotation of the Earth (illus. 9). Moving from science to technology, among the big-and-oily industrial things at the heart of so-called science collections are objects representing the energy industry, such as an oil rig chunk of which I am especially fond (illus. 8). Many technology objects can be much, much smaller, like the strange ferrofluid illustrating inconceivably minute nanotechnology (illus. 10). If this seems exotic, many other objects are more mundane, such as a typewriter illustrating a form that was popular a century ago before the dominance of QWERTY keyboards (illus. 11). Two 'OncoMice' in the Science Museum in London, which are taxidermic remains of two genetically modified rodents, provide tiny exemplars of the museum's strength in its controversial biomedical collections. Generally, however, healthcare collections comprise things like surgical instruments and drug jars. Finally (for now), the Lanz Bulldog tractor (illus. 13) encapsulates both agriculture and transport, which take up a considerable part of the volume of science collections. We will meet other objects along the way in various degrees of detail, but these half-dozen mascots, and those who curate them, will be our guides through this book.

Together they provide a snapshot of the variations across collections. Science museum objects can be very old or very new, from the pendulum to the ferrofluid. 'You owe me a new mobile,' grumbled a colleague from another department, soon after I began working in the National Museums Scotland's Science and Technology curatorial team. His daughter had seen her iPhone model (a recent one) in a new gallery of the museum, and had concluded that it was therefore historical and should be replaced. Never mind that the exhibit was showing contemporary technology, in her mind the museum was indelibly associated with bygones. Even though there are some exquisite objects from the early modern period, a fetish for the first industrial revolution and a growing number of recent artefacts, nineteenth- and especially twentieth-century things dominate. Surveying more than one

10 Tiny science. Magnetic nanoscale particles suspended in a solvent – 'ferrofluid' – under the influence of a strong magnetic field, used by the National Informal STEM Education (NISE) network.

11 Everyday technology that has since become unusual: an early form of typewriter, the Mignon 3, at National Museums Scotland.

hundred star items from three collections, I found that around half were from the twentieth century, a third from the nineteenth and the small fraction remaining were from other periods.[23]

They vary in size as well as age, clearly. Science museum objects can be tiny, like the nano-particles in the ferrofluid, or giant, like the instrument array from which the copper cavity came. The 'Collider' display at the Science Museum, which featured a similar copper cavity, was, as its curator mused, 'an exhibition featuring technology much too large to fit in a museum, seeking particles much too small to be seen in a museum.'[24] The accelerator was an example of the established practice of bringing in manageable chunks to represent massive systems, and the Deutsches Technikmuseum in Berlin collects slices of roads for this reason (illus. 12). The Bulldog tractor may be moderately sized, but among the largest objects in any museums are engines, whether stationary or for rail transport. Most science museums worth their salt will have an early steam locomotive in their central gallery (see illus. 15), and some feature more recent engines

12 A road surface section from the Deutsches Technikmuseum collection.

13 Science collections often include transport and agriculture. The Deutsches Museum's adorable Lanz 'Bulldog' Tractor.

in all their vast glory (illus. 14), which by their sheer size can become a 'quasi-architectural feature'.[25]

The contrast between the antique steam locomotive *John Bull* and a swathe of interwar Bavarian road surface illustrates well that science museum objects can be rare (like the OncoMice) or commonplace (like typewriters). One of the challenges we face is that many important technical things are utterly mundane –

14 The sleek 1936 Canadian National Railways '6400/u4a' locomotive at Canada Science and Technology Museum, nearly 30 metres (100 ft) in length.

bicycles, for example – but if we are to represent the relationship between science and society we need to collect them and to tell their stories in an engaging way. We need to overcome the challenge of their drab appearance, from the road bricks to the shades of grey and beige predominant in twentieth-century scientific apparatus.[26] Scientists often disregard the commonplace kit in their laboratory as meaningless, when a test tube could be a very powerful way of engaging with their work. Again, the Deutsches Technikmuseum provides a neat example: one of their star objects is a billiard ball, collected as evidence for the importance of plastics to twentieth-century culture.[27]

Both the rarity and size of some techno-scientific objects explain the endemic presence of another category of thing we will find: not original objects at all, but rather models or replicas. It would be quite hard to display DNA, so we display a model that is much larger (illus. 16); aircraft and buildings are rather large so

15 The early steam locomotive *John Bull* in the former Arts and Industries Building of the Smithsonian, 1920; it remains front and centre at the National Museum of American History.

we display models that are smaller; nuclear power stations are dangerous so we display models that are much safer. One might have thought that the preponderance of models would be a problem given that museums are ostensibly places for authentic objects, but actually audience evaluation has shown that visitors are not especially bothered by this, if the replica they encounter is interesting.[28]

To recap: not every science museum object is real; not every science museum object is rare; not every science museum object fits in a case; and not every science museum object is old. Finally, not everything in a museum collection is even an object. In museums generally, and science collections specifically, formally registered objects are in a minority. Museum time, space and work are as much taken with textual, visual and digital entities as they are with the material realm. There are records, images and electronic files associated with material culture but also collections

16 Many science objects are models – of the very small and complex, in this case: an early model of DNA from the Science Museum's collection.

that are valuable in their own right. For example, the Science Museum Group's impressive 425,000 artefacts are numerically dwarfed by more than 7 million books, manuscripts and photographs. And increasingly, there and elsewhere, these are represented and accompanied by digital files. While exploring science collections in what follows, we will therefore encounter not only material things (our main preoccupation), but words, pictures and code.

What Is a Science Curator?

What unites these diverse material and immaterial things is their capacity to tell stories about science? Who is telling these stories? There are many practitioners in and around science museums crafting narratives in different media, but in this volume, although educators, students, conservators and others will feature, we will focus mainly on curators. Which brings us to another important point: a science curator is not a scientist. Whereas natural history curators tend to be scientists, whether palaeontologists, zoologists or entomologists, by contrast even though many science-and-technology curators have a science background, we tend more often to be historians. The nuances of these differences do not always filter through: when the *Scottish Review* selected National Museums Scotland science curator Sophie Goggins as one of their '20 outstanding 20-somethings living and working in Scotland', she was dubbed a scientist.[29] True, she studied science at university, but in her day-to-day work she does not practise science. She is a curator. What, then, does this involve?

The term 'curator' has a broad cultural presence, associated with anything that involves creative selection.[30] Festivals are curated. Menus are curated. Playlists are curated. Which is all well and good, but it does mean that we might need to think more carefully about what a museum curator does. A time-and-motion analysis of a curatorial role in the twenty-first century would reveal hours on email and in meetings like many other professionals, but they also spend time in museum galleries and collection stores. Generally, their duties would be divided between three elements around which the later part of this book is arranged: collections, research and engagement. The former includes acquiring new material and caring for existing collections. Curators then undertake their own research on these collections, collaborate with others and/or respond to queries. This collecting and research then gives rise to exhibitions, events and digital activity. Plenty of administration underpins these core functions, and they vary over time and organization.

Training varies too: there is no set route, but for the most part curators will have at least a graduate and possibly a postgraduate degree. Many will have studied science and/or museum studies, or sometimes the history of science or history of technology. Almost all will have volunteered with a collection at some point – and through training and their everyday work, directly or otherwise, there is a connection to the objects running throughout. 'Curator', after all, comes from *curare*, to care.

It is helpful, too, to recognize that most museum staff are not curators. There are many roles in larger museums that are also concerned closely with collections: conservators who care for them; registrars who manage their movements and information; exhibition makers who display them; learning and digital experts who elucidate them; and front-of-house staff who help visitors to access them every day. Museums are multi-service institutions, with cleaners, fundraisers, shop assistants, chefs, porters, public relations managers and more. Curators may be in a minority, but they are nonetheless the protagonists in this book, and demystifying curatorship is our principal focus.

These various training routes, the different organizations and characters, mean that there is as great a variety among science curators, as there is in any curatorial community clustered around a particular kind of collection. What unites them is a passion for the kinds of things that are the subject of this book. They have powerful emotional attachments to the collections in their care. 'You couldn't do this job in a very mechanical way,' one curator reflected, but 'you do actually end up loving your objects. And you do end up thinking of them, however much you know you shouldn't[,] as *your* objects.'[31] At worst, this instils a Gollum-like protectiveness and an unhealthy territorialism; at best, it makes a good curator a passionate storyteller. Ultimately, stories are our job: when documenting things, when finding out more about them, when using them in exhibitions and online. When I first arrived at National Museums Scotland, I asked the science and technology curators what we should be collecting. 'Stories,' they unanimously replied. This book is about how we tell those stories.

Who Visits Science Museums?

Storytellers need an audience. Curators undertake these activities for the benefit of museum visitors and other users, whether now or in the future. Not all of them are children, despite science museums' reputation. A common response to science and technology galleries in museums is 'Ah – my children love that section!' To many, science museums are exclusively associated with manic, noisy, push-button bustle. 'There is, apparently, an ideal level of energy at which children learn about science,' observed one weary journalist: 'Too much giddy excitement and information spins off their flywheel minds; too much sober exposition and boredom numbs them.'[32]

Whatever their age, there are certainly plenty of visitors. In the UK, around a fifth of the population visit a science museum or science centre each year (admittedly compared to a third for natural history and around the same for art museums).[33] The 'Nano' travelling exhibit featuring things like the ferrofluid (see illus. 10) attracted up to 11 million people per year across the different sites in North America; the larger science museum institutions boast this order of magnitude on their own. The Air and Space Museum in Washington and the Science Museum Group in the UK both attract upwards of 6 million visits a year. A national museum like the Museo della Scienza e della Tecnologia can boast over half a million visits per year; the Polytechnic Museum in Moscow is closed at the time of writing, but its temporary exhibition in the meantime has more than 600,000 visits annually; a specialist but well-positioned science collection, like the MIT Museum, attracts around 150,000 visits per year; a more focused university collection, like Medical Museion in Copenhagen, has a footfall of some 30,000. Most economically viable science centres, meanwhile, will need over 100,000 to survive.[34] Together, it is likely that science museums and their ilk attract over 120 million visits per year globally; and these numbers are likely to be matched by online visitors.[35]

Who are they, these millions, given they cannot *all* be children? Science museums have always had a dual appeal: attracting researchers and other elite audiences while seeking to attract the workers and the general public for edification. But who comprises this 'general public'? It is difficult to generalize about these millions, but the Wellcome Trust has shown that the likelihood of visiting a science museum or centre is related to social class: those in professions and their families are three times more likely to visit than the long-term unemployed and those working in manual occupations.[36] Museum audiences are not as diverse as the wider population: Black people, for example, are under-represented in their visitor constituency (as well as being too-often absent in their displays).[37] Geography is critical, because science museums, like other cultural venues, seek to position themselves on the international tourist trail. The MIT Museum, for example, attracts almost a third of its visits from overseas visitors, and this is not uncommon in major cities.[38]

These people are of different ages. Adults have not, as one historian has disingenuously argued, disappeared from science museums.[39] Besides the specialists we shall meet ferreting around behind the scenes, there are plenty of grown-ups wandering the exhibits. In any case, to think about this in terms of adults versus children is to miss the point. Science museums have an unusually heterogeneous constituency in terms of age. Science objects facilitate intergenerational interaction, sometimes pedagogical, sometimes purely social. Visitors talk to each other, learn from each other, have fun with each other. And while we might immediately think about the 'adult–child dyad', other interactions are just as important: between adults, among children. One important visitor configuration, for example, is of grandparents or other older relatives visiting with young people. Some of these adults may be enthusiasts about elements of the collections on display, keen to infect their young charges with the passion they enjoy. They may not be successful, and neither will other permutations of expertise and interest within a group. Adults may be nervous of their own expertise, and they, like their children, may get bored – but this interaction is nonetheless key.[40] A museum visit is not only a visual and textual experience, but a social one.

Exploring Science Museums

If you have picked up this book, you may well be one of these visitors or a science museum 'user'. If you know a little bit and you have a curiosity to learn more, this book is for you. If you are a fellow science museum junkie already, this book is for you. If you are a scientist or a historian of science with an interest in material culture, this book is for you. If you are fortunate enough to work in a museum, or you are thinking about it, studying or training to do so, this book is for you. I suspect that those in the latter categories will not agree with everything I say – certainly I hope not – and those in the former should be aware that we do not all agree. What follows is a brazenly partial take on the function of science collections.

As such it is intended to be a contribution to an emerging genre that we might call 'curatorial confessional', of which palaeontologist Richard Fortey's *Dry Store Room No. 1*, about his time at the Natural History Museum in London, is the most famous.[41] He and other authors of his ilk love collections and they want to take their readers behind the scenes. At best they offer social histories of museums from the inside, interweaving memoir, critique and an element of 'museology' (the ill-defined academic field that is an admixture of museum theory, reflection and policy). At worst they churn out self-indulgent exercises in score-settling and grumbling about how 'things ain't what they used to be'. I hope what follows is more of the former than the latter. As a study of a particular kind of collection, this is also my contribution to the museology of science: there have been some excellent studies in this vein, but not for a while.[42]

This book is intended, then, to be more personal polemic than academic analysis. My approach to supporting my arguments has been selective rather than systematic. Just as I have personally selected the copper cavity and my other favourite objects, so I have cherry-picked from the plentiful writing about collections generally. There is a growing library of analytical literature and a bevy of periodicals devoted to these topics (you can follow the endnotes and browse the Select Bibliography if you are interested in learning more). I have carried out 'fieldwork' in museums, visiting curators, exhibitions and storerooms; I have interviewed and observed professionals and users; and I have added a dose of my own experience and observations. This makes for a blend of synthesis, autobiography and ethnography with a touch of commentary. If I were feeling grand, I might call it 'auto-journo-ethnography'.

Whatever it is called, my aim in unlocking European and North American science museums is to understand how objects like the copper cavity have been, are and should be used, and by whom. Along the way, we will unpack some of the assumptions that curators and stakeholders make about these objects. The tasks of collecting and exhibiting science objects are difficult, expensive and often thankless. Given there are many other ways of recording and experiencing science – not only physically, but digitally and televisually – why do we continue assembling physical collections? What should the function of the science collection be in the twenty-first century? To answer these questions, we will travel together through science collections, past, present and future. We embark with how they came to be (Chapter One), then how they are collected (Chapter Two), followed by their use in research (Chapter Three), engagement (Chapter Four) and advocacy (Chapter Five). In each case we will unlock assumptions or paradoxes about science collections, and use them as entrées to that museum function. In each we will encounter good practice (and some bad), current challenges and indications as to what science museums *should* do, which is the particular focus of Chapter Five.

This introduction has already provided a taster of some of the surprises that we will uncover along the way: that science collections are more about people than principles; as much about the now as the then; and as much about the intangible as the tangible. Science collections are full of contradictions like these. They are flexible enough to hold the antique and the cutting-edge; they show great discoveries and unfinished research; they appeal to schoolchildren and Nobel Prize winners; they accommodate the massive and the minute; they are geared towards the physical but seek to represent the intangible; they show geographically specific evidence of a de-localized enterprise; and they are associated with object-free interactivity while storing hundreds of thousands of resolutely material things.

The door to the science museum is unlocked. Please, be my guest.

1

HOW COLLECTIONS CAME TO BE

The 20-centimetre (8 in.) brass-covered lead globe shown here (illus. 17) is suspended from a pivot high above; the axis of its swing gradually rotates over the course of a day, demonstrating the rotation of the Earth. However, as all images do, this picture misses both its brute physicality – it weighs 30 kilograms (66 lb) – and its movement. This one now swings back and forth in the former abbey of Saint-Martin-des-Champs in Paris, greeting visitors to the Musée des arts et métiers, one of the oldest and grandest science museums in the world. The pendulum is an ongoing, dynamic demonstration of a scientific principle. Its hybrid nature – at once dynamic experiment and scientific antique – is common to science collections more broadly.

Every day curators at the Harvard University Collection of Historic Scientific Instruments walk past a humble trolley of boxes (illus. 18 and 19) that came from the collection's founder curator David Pingree Wheatland. He first worked in his family's lumber business in Maine, but his passion was for collecting scientific apparatus, for which he acted as curator at Harvard for a nominal $1 per year. His professional roots emerged in quiet ways through his collection: in this case he used a stock of boxes from the family business to store electrical parts. As a way of keeping visual track,

17 Foucault's pendulum in the Musée des arts et métiers, formerly the Saint-Martin-des-Champs, Paris.

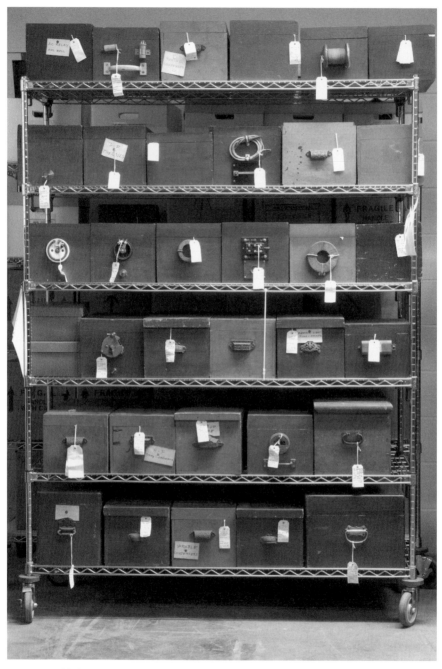

18 A trolley of electrical equipment parts in the Harvard University Collection of Historic Scientific Instruments. Founder David Wheatland (1898–1993) stored them in boxes from his family business, with an example of the part in the box displayed on the outside.

he attached an example of the objects inside to the outside of the box: rheostats, coils, cables. Decades later, collection staff still use them fondly. This simple organizing device gently demonstrates that efforts to collect, order and display the world of science are contingent on the people involved at a given institution at a given time. Their improvisation and their idiosyncrasies endure, even as the institutions they founded go from strength to strength.

Science museums present universal truths (like the movement of the Earth) based on vast collections. And yet they are each particular: to their place, their past, their politics, and the people involved. In this chapter we see the history of science museums not as a grand march of progress, but rather a far more interesting, political and above all human tale of particular people at particular times. Some collectors founded and worked in small collections like that at Harvard, others at grand institutions like the Musée des arts et métiers in Paris, the Science Museum in

19 One of Harvard curator David Wheatland's repurposed storage boxes, showing its previous life filing vouchers – presumably from the Maine Employment Security Commission.

London, the Deutsches Museum in Munich and the Exploratorium in San Francisco. In this chapter we will meet their weird and wonderful founders: a revolutionary clergyman, an indefatigable technocrat, a 'superintendent of specifications', a social-climbing engineer and a disgraced scientist. Science museums are the result of their quirks and passions, of serendipity and happenstance, as much as they are of strategy and planning. Museums deal in what is available and what survives; even these grand collections are local and often political manifestations of an ostensibly universal enterprise.

These idiosyncrasies run through the twin histories of science collections: a history of expositions and interactivity on the one hand and a tale of antiquarianism and historical scholarship on the other. Throughout their lives, science museums have balanced past and present; celebration and interaction; brass-and-glass and push-button; connoisseurs and kids. These tensions are not new, and neither are these institutions. Even though major players like the Science Museum and the Deutsches Museum were officially founded around a century ago alongside other grand imperial natural history and art museums, we will find that the roots of their collections are much deeper. They begin in Renaissance cabinets of curiosity, enhanced during the Enlightenment and then expanded by international expositions in the nineteenth century. Dedicated science museums were already thriving by the time of their significant expansion in the early twentieth century; and although the Second World War paused their growth, they continued to flourish in the post-war decades. The science centres that joined traditional science museums during the twentieth century were only the latest manifestation of hands-on offers for visitors, and the debates around the millennium about the best way to engage with these audiences had a long, long history.[1]

Let us look over this history, first with a wide focus on the long roots of collections, then in more detail at the development of science museums as we now know them. At first we will consider a wide range, and that range flourishes to this day, including dozens, even hundreds, of private, specialized and idiosyncratic collections.[2] But for our present purposes, for the more recent history we will focus on a few key institutions – including, counter-intuitively in this book about collections, recent museums like the Exploratorium that do not appear to have collections at all.

Spectacular Collections

It turns out there were science collections before there was science. Among the diverse princely cabinets of art from the sixteenth century, before anyone had got around to inventing 'science', could be found instruments for surveying, measuring, optics and mathematics. What linked these exquisite instruments with the other (to our eyes) miscellaneous items was their *artfulness*: it was in this sense that these were cabinets of the arts.[3] For example, one of the fabulously wealthy Medici family in Florence, Grand Duke Cosimo I (1519–1574), collected mathematical instruments and housed them in his palace; and later the French aristocrat Joseph Bonnier, Baron de la Mosson (1702–1744), included mathematical devices in his extensive cabinet (illus. 20).

While other elements of these multifarious cabinets would go on to be included in art, anthropology and natural history collections, by the early eighteenth century many of the instruments found their way into more focused collections such as the Mathematisch-Physikalischer Salon in Dresden.[4] These instruments were intended not only to enhance the esteem of their cultured patrons, but for them to use, to demonstrate the new 'natural philosophy'. Innovative electrostatic machines and air pumps in collections were hands-on, interactive devices used to stage demonstrations of what we would now call science (illus. 21).[5] Generally, witnesses of these demonstrations of early modern collections were other members of the elite. Even the supposedly public British Museum, which opened in London in 1759 and included some scientific instruments alongside other 'natural and artificial rarities', was only accessible to the well-connected.[6]

An approach we would now consider a more democratic approach to audiences today could be found in a new institution that opened in revolutionary Paris at the end of the century. Henri Grégoire was a Catholic bishop and a leading member of the National Convention that ruled France in the wake of the bloodshed of 1789. For him, the end of the *ancien régime* was a step towards universal suffrage; he argued for racial equality and the abolition of slavery, advocated Jewish emancipation and supported the new state of Haiti. Grégoire also wanted to give working men the opportunity to better themselves. Even when the revolutionaries turned against the clergy, he continued to wear his garb and tried to protect libraries and artworks. He retained his position on the ruling convention and was instrumental in founding the organization that gave rise to the Musée des arts et métiers in 1794 to improve national industry and showcase revolutionary progress for the workers. It was housed, appropriately given its clerical driving force, in a former priory. 'There will be formed in Paris', he proclaimed, 'a depository for machine, models, tools,

20 Princely cabinets included scientific instruments: Jacques de Lajoue, *Interior of the Physics Cabinet of Bonnier de la Mosson*, 1734, oil on canvas.

21 Demonstrative science: Joseph Wright 'of Derby', *An Experiment on a Bird in the Air Pump*, 1768, oil on canvas. Presented to the National Gallery by Edward Tyrrell, 1863.

drawings, descriptions and books in all the areas of the arts and trades.' Among these 'arts and trades' are things we would now classify as science. The exhibits on display in Saint-Martin-des-Champs included state-of-the-art machines and instruments, useful and instructive, for workers to see and use. Experienced craftsmen were to be on hand to demonstrate and explain. For Grégoire, museums were 'workshops of the human mind'.[7]

As the Industrial Revolution gathered pace, other institutions followed suit in establishing displays to inspire and inform the workers. Among the museums of different kinds emerging in Philadelphia, another city with the fresh whiff of revolution, industrialist Samuel Vaughan Merrick and geologist William H. Keating set up the 'Franklin Institute of the State of Pennsylvania for the Promotion of the Mechanic Arts'. They staged their first exhibition in Carpenter's Hall on 'American Manufacturers'. The organization survives to this day, now as the (rather snappier) Franklin Institute.

In London the dazzling exhibition ecosystem included the Adelaide Gallery (founded in 1832) and the Polytechnic Institution (1838), both offering lively, hands-on spaces for 'practical science'.[8]

It is not clear how far places like these succeeded in piquing the interest of workers enhancing industrial output. The devices on show were necessarily the shiniest and the best, but were not always replaced as they aged. They became venerable rather than instructional; but no less appealing for that. The Musée des arts et métiers, which from the outset displayed historical material next to recent developments, also absorbed the princely instruments of the Academy of Sciences, which were more awe-inspiring than interactive. The collection would attract such technical *Mona Lisas* as Blaise Pascal's calculating machines, Antoine Lavoisier's laboratory equipment and the Lumière brothers' camera.[9]

Among these iconic artefacts, the Museé des arts et métiers also housed one of the earliest examples of the pendulum designed by nineteenth-century French physicist Léon Foucault (see illus. 17). Having demonstrated the first version in 1851 at the Paris Observatory and then under the dome of the Panthéon (where a new version was installed in 1995), four years later he demonstrated an iron version in the Palace of Industry at the universal exposition. Both were acquired by the Musée in 1869, and many other large science museums followed suit in due course, typically hanging a pendulum in their atrium or stairwell.

This acquisition helpfully illuminates the importance of the relationship between such 'expos' and museums in the nineteenth century. Paris had hosted a series of industrial expositions throughout the early nineteenth century, and whatever the French could do, the British set out to do better. Their response was developed by Henry Cole, a civil servant who was deeply interested in industrial design and was active in the Royal Society for the Encouragement of Arts, Manufactures and Commerce, where he found a key ally in Prince Albert. With the prince consort's support, Cole arranged a series of exhibitions showcasing the latest in British design. Spurred by a visit to the 1849 Paris exposition, Cole with boundless energy pressed for a truly international exhibition – *The Works of Industry of All Nations* – which Albert and Queen Victoria opened in London in 1851. The latest scientific instruments, the grand 'Machinery Court' and other material manifestations of *progress* were well represented among 13,000 exhibits illustrating art, nature and culture. It was successful, attracting 6 million visits and generating a significant surplus.[10]

In the wake of the exhibition, the British Government decreed that scientific education in Britain needed improvement and established the Department of Science and Art with Cole at the helm alongside the ambitious Scottish scientist Lyon Playfair, who had also been closely involved in the Great Exhibition. Using scientific (and other) collections at the heart of an intended system of technical education, they established a museum based on the leftover collections of the Great Exhibition, including both industrial and decorative arts; to which they added more technical apparatus, such as ship models. The South Kensington Museum, incorporating the Museum of Manufactures, opened in makeshift buildings (the 'Brompton Boilers') in 1857. It encompassed material from across the useful and decorative arts, giving rise not only to the Science Museum but the Victoria and Albert Museum.[11]

Cole also wanted to set up museums elsewhere in Britain, and set his eye upon Edinburgh. There he found fertile ground; Lyon Playfair, on his way to take up the chair of chemistry at the university, and fellow professor George Wilson had already gathered a collection of 10,000 machines, models and samples known as the Industrial Museum of Scotland. This was absorbed into the Edinburgh Museum of Science and Art in 1866, next to the university (whose natural history collections it incorporated), in the working-class Old Town, demonstrating with artefacts and working models the latest in science and industry for the workers of Scotland.[12] The Department of Science and Art also set about stimulating the design and manufacture of scientific apparatus. To showcase the latest available, in 1876 South Kensington hosted the display of 20,000 scientific instruments, the 'Loan Collection', that would form the basis of the instrument collections in what is now the Science Museum (illus. 22).[13] It was a massive exercise in scientific bravado, masking a growing fear that Britain was losing its place at the forefront of science and industry.

The race was on. The vogue for grand expositions gathered around the world, many following the South Kensington model of going on to be the core of long-standing collections. Such 'World's Fairs' that gave rise to permanent museums included the Intercolonial Exhibition of Australasia in Melbourne in 1866–7, the 1873 Vienna Universal Exposition, the Centennial Exhibition in Philadelphia in 1876, the 1888 Glasgow International Exhibition and the 1914 Jubilee Exhibition in Oslo. By the time Paris held the 1900 international exposition there were more than 83,000 exhibitors and nearly 60 million visits (ten times that of the Great Exhibition).[14] They spanned arts and culture, but for our present purposes we should note the instruments, inventions and machines they included that remain in museums near and far to this day, partly thanks to savvy requests for objects made by curators and museum founders when the expositions concluded. The expos and the museums they begat were intended to boast the national expertise of the exhibitors, further their imperial ambitions and to inspire the working people of the host nation. This was cutting-edge science, the technologies of today and tomorrow.

Further places to encounter groups of innovative things were collections associated with national patent offices. Many new inventions were accompanied by models or prototypes, which were not only lodged for record but intended to inspire the next generation of inventors. The British Patent Office Museum crystallized in the 1850s under Bennet Woodcroft, previously the University of London's professor of machinery and then the Patent Office's gloriously titled 'superintendent of specifications'. It eventually transferred to the Department of Science and Art and merged with the museum in South Kensington; the technology elements of the present-day Science Museum have their roots in this material. By parity the United States Patent Office in the early twentieth century transferred its vast collection of working models to the multidisciplinary Smithsonian Institution, America's national museum, which had been founded in 1846.[15]

J.T. BALCOMB, DEL.

1. Tycho Brahe's quadrant.
2. Sir Francis Drake's astrolabe.
3. Galileo's telescope.
4. Galileo's second telescope.
5. Newton's telescope.
6. Janssen's compound microscope, 1590.

7. Galileo's microscope (occhialini).
8. Sir Humphrey Davy's first safety-lamp.
9. Third safety-lamp.
10. Davy's improved safety-lamp.
11. Pascal's adding and subtracting machine, 1642.

12. The "Napier Bones," for divide and multiplication, about 1700.
13. Sömmering's electric telegraph, 1809.
14. Faraday's magneto-electric induction apparatus.
15 and 16. Faraday's later apparatus.
17. Forbes's apparatus.

18. Galileo's air thermometer.
19. Dalton's mountain barometer.
20. Dalton's apparatus for testing the tension of ether vapour.
21. Ancient Swiss clock, from Dover Castle.

HISTORICAL TREASURES IN THE LOAN COLLECTION OF SCIENTIFIC APPARATUS, SOUTH KENSINGTON.

22 Historical treasures in the Loan Collection of Scientific Apparatus, South Kensington, engraving by J. T. Balcomb in the *Illustrated London News*, 16 September 1876.

The patent material joined a significant collection of now-historic scientific apparatus already gathered by the Smithsonian's first secretary, Joseph Henry, as well as an ageing collection of scientific instruments from the United States Centennial Exhibition in 1876.[16] Although patents and expositions necessarily focused on contemporary instruments, curators also collected and displayed scientific equipment for historic purposes throughout the nineteenth century. In London, Bennet Woodcroft was fascinated by historical machinery, spending considerable energy seeking out 'relics'. Stephenson's historic locomotives, *Rocket* and *Puffing Billy*, were displayed at the Loan Collection exhibition and he acquired them for the patent museum. Even the 1876 cutting-edge display included 'not only modern apparatus but also apparatus interesting from the persons by whom it had been employed or the discoveries in which it had been used', including instruments used by Galileo, Lavoisier and Joule – the 'sacred relics' of science.[17] They were important for nation-building, for demonstrating the pedigree of science. Elsewhere, as at the Museé des arts et métiers, scientific instruments and technical artefacts collected as the latest innovation remained in collections and became historical by the simple passage of time. Institutions like the Teylers Museum in the Netherlands began as active research spaces but then acquired a heritage function almost by accident.[18] One might consider them fossilized; but in the kinder words of one curator, 'they discovered the virtues of history.'[19]

It seems, then, that the twin identities of science collections – present versus past, use versus veneration – go back as far as we care to look. But the sheer scale on which we experience them today was a twentieth-century development.

Large-Scale Collections

One visitor to the first International Exposition of Electricity in Paris in 1881 took it to heart. Young engineer Oskar von Miller, scion of a notable Bavarian family, was short neither of ambition nor of patriotism. Infected with the exhibiting bug, he returned to Munich and set up a German equivalent of the Paris expo. In concert with a successful engineering management career, he set about collecting for a museum fit for a unifying nation flexing its industrial and scientific muscles. Von Miller's particular skill was in garnering support for his museum idea: from scientists, engineering associations, companies, wealthy benefactors – and, like Henry Cole before him, even from royalty.

It took some time, but by 1903, at the height of the Second Reich, he had formally established the Deutsches Museum von Meisterwerken der Naturwissenschaft und Technik (the German Museum of Masterpieces of Science and Technology).[20] The city of Munich gave over an island in the river Isar, which had appropriately been known as Kohleninsel, 'Coal Island', because charcoal had been stored there. Money came from Bavaria and from the imperial government – museums of all kinds were tools of empire – and Kaiser Wilhelm II laid the first stone in 1906. Delays and war intervened, however, and its own building did not fully open until von Miller's seventieth birthday in 1925. The long gestation meant that many of the artefacts in its 30,000-square-metre (322,920 sq. ft) exhibit space had, in the meantime, become historical, complementing in heritage terms the relics and celebrity apparatus. In the Ehrensaal (Hall of Honour) portraits of famous German scientists and their equipment were displayed as masterpieces of science and technology.[21] There were also plenty of cutting-edge science demonstrations and interactive experiences; this was the first time push-button and hand-cranked working models were displayed on such a massive scale. And in this dual function, the Deutsches Museum was the best of its kind in the world.

It had some serious competitors, however. The technical collections in the South Kensington Museum, effectively separate since the 1880s, had been formally dubbed the Science Museum in June 1909, proudly rubbing shoulders with its sister collection at the Victoria and Albert Museum, and the Natural History Museum next door. (Getting in early on the international stage meant the British had no need to specify the nationality of *the* Science Museum.) Combining the technical elements of the Great Exhibition

23 Young Science Museum visitors in 1951 getting hands-on with one of the original exhibits in the Children's Gallery.

with the Patent Museum, many of the treasures of the 1876 Loan Exhibition, and later material from the 1924–5 British Empire Exhibition (vivid evidence of the imperial context of these organizations), the collection was then arranged in three divisions: marine engineering, machinery and inventions, and scientific apparatus. Each included both historic treasures and recent developments. Tellingly, control of the museum had transferred to the Department of Education when the Department of Science and Art disbanded in 1899; technical education was the principal function of the collection. There were similar developments at the former Edinburgh Museum of Science and Art, by this time the Royal Scottish Museum, where its Department of Technology inherited the collection seeded by George Wilson, displayed in halls devoted respectively to power, mining and shipping. A 'Science Gallery' opened at the beginning of the twentieth century showcasing apparatus of renowned Scottish scientists.[22]

Elsewhere, technical museums such as those founded in Prague (1908), Oslo (1914) and Vienna (1918) proclaimed both the heritage and the current ingenuity of their nations, as well as their imperial aspirations in many cases. Working with industry to secure hot-off-the-press devices, they juxtaposed working exhibits with sanctified relics, the evolution and operation of national science. Philanthropist Julius Rosenwald was inspired by a visit to the nascent Deutsches Museum to fund the Chicago Museum of Science and Industry, which opened during the 1933 Century of Progress Exposition, housed in the grand Palace of Fine Arts from the World's Columbian Exposition four decades earlier.[23] It focused on scientific principles and recent discoveries and set out to make the exhibits animated. Visitors pushed buttons and looked through microscopes. One such visitor, Alexander Hutchieson, keeper of technology at the Royal Scottish Museum, was impressed. He returned to Edinburgh to oversee a workshop dedicated to building painstaking replicas of engines and other machines, which operated in the galleries to the delight of visitors.[24]

These were lively places. Oskar von Miller himself had visited the United States several times and at a dinner in the 'Museum of the Peaceful Arts' in New York in 1929, he encouraged active participation in science exhibitions. Curators from several u.s. museums were present and the word spread. In 1934 the venerable Franklin Institute opened a hands-on science museum as a 'wonderland of science' explicitly modelled on the Deutsches Museum. So too did the New York Museum of Science and Industry, which took on the Peaceful Arts' collections. Its opening in 1936 included a speech by Sir William Bragg, by telephone from the Royal Institution in London:

> he gave a short address to a distinguished gathering including Prof. Albert Einstein . . . listeners then heard Sir William strike a match, with which he lit an old candle set in a candle-stick of [Michael] Faraday's time; in a few instants, the entrance hall of the New York Museum was flooded with the light of two rows of mercury vapour lamps.[25]

The transatlantic electrical impulse that triggered the floodlights was at once an illustration of photo-electricity, visual evidence of American ingenuity (the signal passed through American telegraph and telephone lines) and a theatrical flourish. This approach was mirrored at the new Palais de la découverte in the Grand Palais, Paris, inspired by the Nobel Prize-winner Jean Perrin, who wanted to demonstrate laboratory science at work. It was in effect a forerunner of science centres later in the century and, importantly, this dynamism attracted more and more children to the museums. In 1931 the Science Museum responded to its million visitors per annum by opening a new gallery explicitly for young visitors (illus. 23). With working models and hands-on experiences adult visitors also found it fun. In Edinburgh a young visitor later recalled, 'the thing that attracted us was whether we could press the buttons – but of course the trouble was [every]

24 The University of Oxford's Museum of the History of Science (now the History of Science Museum) in 1951.

25 Historian of science I. Bernard Cohen and curator David Wheatland with an orrery at Harvard University, *c.* 1947.

little boy wanted to press the buttons, you couldn't always get in the queue quick enough.'[26]

And yet young visitors, whizz-bang interactivity and the lure of the new did not rule out the heritage function of science museums. Collectors and antiquarians were increasingly attracted to historic scientific instruments in the interwar period as interest in the history of science increased; this was reflected not only in established science museums but in a new breed of private collections. Avid connoisseurs amassed collections of unprecedented size that they exhibited proudly. They included Julius Rosenwald's equally wealthy brother-in-law Max Adler in Chicago; Robert

Whipple, managing director of the Cambridge Scientific Instrument Company; and near London the paper manufacturer Lewis Evans. These collections then found their way into, or became the foundation of, museums: Adler deposited an expensive collection of historic instruments he had purchased in the planetarium he opened in 1930; the Museum of the History of Science in Oxford crystallized around Evans's material in 1935 (illus. 24); and the University of Cambridge opened a museum for Whipple's collection in 1944. Universities were fertile grounds for presenting the pedigree of science: instrument collections that had been used for teaching in science departments often aged and acquired heritage values, illustrating the history of their own institutions' innovative contributions to science, past and present, which could be writ large in objects. Harvard is a good example of this: David Wheatland began collecting instruments in the 1930s that would be the core of the museum there (illus. 25). On a national level, against the background of burgeoning museums of all kinds, collections trumpeting deep national scientific heritage included the new Istituto e Museo di Storia della Scienza in Florence (established in 1927, now the Museo Galileo) and Het Nederlandsch Historisch Natuurwetenschappelijk Museum in Leiden (established in 1931, now the Rijksmuseum Boerhaave).[27]

Reinvigorated Collections

Whether historical or hands-on, resources for science collections were understandably slim during the Second World War. Some premises were used for other purposes (the Museum of the History of Science in Oxford was used as a labour exchange, the Royal Scottish Museum a storehouse for hospital supplies) and collections taken away from city centres for safe-keeping. Stephenson's *Rocket* left the Science Museum for Brocket Hall in Hertfordshire (later the setting for actor Colin Firth to emerge dripping from a lake in the BBC's 1995 version of *Pride and Prejudice*).

Museums of all kinds began to dust themselves off in the years following the war, and science collections were among them. The Boston Museum of Science opened in a new building in 1951. Two years later the Museo della Scienza e della Tecnologia Leonardo da Vinci opened in a repurposed bomb-damaged monastery in Milan. In Britain, the optimistic post-war government planned a grand exhibition to mark the centenary of the Great Exhibition and a sign of post-war recovery. The Science Museum hosted an exhibition in its still-not-complete centre block. Here and at other venues such as Kelvin Hall in Glasgow, science and technology were at the forefront of this 'Festival of Britain'.[28]

In the West, both this buoyant mood and the significance of science museums changed overnight on 4 October 1957 when the Soviet Union launched Sputnik I, the first artificial satellite. Six weeks later, the u.s. government overhauled science education, giving science museums a keen patriotic mission to enhance the 'science base' of its citizenry and to emphasize national progress. During the decades that followed, science museums were agents in the Cold War, educating citizens and emphasizing national achievements, for example via the formal National Science Foundation's Office of Science Exhibits. Museums were to be stages to propagandize the merits of the technologies that powered and were powered by the paranoia of the Cold War, especially around nuclear power.[29]

It was in this context that a new museum emerged within the already sprawling Smithsonian Institution that included one of the most extensive science and technology collections in the world. Already in the 1920s technology curator Carl Mitman had chafed for a 'National Museum of Engineering and Industry', along the lines of the Deutsches Museum.[30] Although this did not materialize, Mitman lived long enough to see his sometime apprentice and long-serving engineering curator Frank Taylor take advantage of the Cold War patriotism to begin a new building on the National Mall; it was finally opened in 1964 as the Museum of History and Technology, its giant ground floor dominated by technology (atomic energy, automobiles, agriculture). Visitors were welcomed by a Foucault's pendulum (illus. 26), then entered galleries in which American technological superiority was clear. 'I hope every school child who visits this capital,' said President Lyndon Johnson at its opening, 'every foreign visitor who comes to this first city, and every doubter who hesitates before the onrush of tomorrow, will spend time in this museum.'[31] The 'Great Marble Shrine' was the first modern building on the Mall and opened in the same year as the New York World's Fair, with which it shared common themes of technological progress and Soviet distrust.[32] Taylor intended it to be 'a permanent exposition that commemorates our heritage of freedom and highlights the basic elements of our way of life'.[33]

The opening of the Museum of History and Technology marked the beginning of a decade of expansion of European and North American museums of all kinds. The public flocked to science collections as post-war rebuilds were finally completed: once again a million visited in London per annum, nearly 3 million in Chicago, and a huge 5 million in Washington.[34] Curators tried to make modern science as exciting and interactive as possible: grandiose electrical demonstrations were popular, especially in the Museum of Science in Boston, which had taken a two-storey high-voltage electric generator from MIT in 1956. Likewise in Edinburgh, the technology galleries remained multi-sensory, popular for their push-button working machines, and even had 'talking labels'. Collections on display were dynamic and the science and technology they represented were forces for good, building towards a utopian future.[35] At the apogee of both this technological utopianism and the visible influence of the Cold War were the lunar-related exhibitions in the 1960s. From the early 1960s world tour of John Glenn's capsule *Freedom 7* (illus. 27) and culminating in the Moon rock exhibitions from *Apollo 11*, spacecraft and anything lunar drew unprecedented crowds.[36] President Nixon gifted the Science Museum a Moon rock wrapped in the Union Flag. This vogue fuelled plans for a dedicated Air and Space Museum in Washington, DC, which opened on the Mall in 1976, the ultimate monument to American Cold War bravado.

The expansion and popularity of science museums in the 1960s and '70s was propelled not only by this neophilia, but by a

26 Visitors on the lower level of the National Museum of History and Technology, now the National Museum of American History, watching Foucault's pendulum, *c.* 1970.

refreshed interest in history that was evident across the heritage sector. For example, in a guidebook from the Science Museum at this time only five of fifty key objects originated after the First World War: the bulk were Victorian, including a Boulton and Watt beam engine and a Trevithick high-pressure boiler.[37] They especially represented the heavy industries that were declining at the time. Even as the 'White Heat of Technology' drove British culture forward a heightened interest in industrial heritage kept an eye on the rear-view mirror. New museums and new posts within existing institutions were established across former industrial areas, many of them *in situ* in former industrial sites.[38] In Manchester an academic history of science and technology group was interested in local industry and with the city council

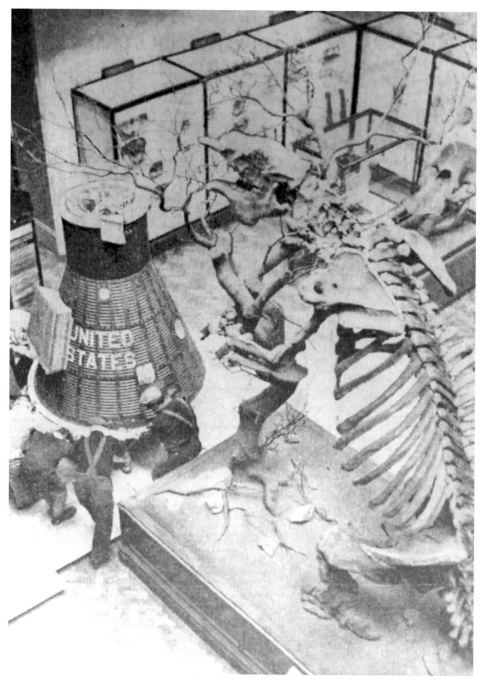

27 Ancient and modern: *Freedom 7* on its way to display at the Royal Scottish Museum (now the National Museum of Scotland), as reported in the *Glasgow Herald*, 13 May 1966.

opened a science and industry museum in 1969.[39] It moved to the former Liverpool Road Station (the first passenger terminal in the world) as the Greater Manchester Museum of Science and Industry in 1983, adding the adjacent Air and Space Museum two years later. Drawing explicit inspiration from the Deutsches Museum, it is now known simply as the Science and Industry Museum (and since 2012 it has been part of the Science Museum Group). Like its industrial peers, including Ironbridge (1967) and Beamish (1970), a visit to Liverpool Road Station was a hands-on, dynamic experience, with working exhibits and artefacts displayed in an industrial setting (even if the objects and the buildings were not always connected).

Whizz-Bang Collections

In the United States there emerged at this time a different way to experience science. There were new kids on the interactive block. Although they ostensibly eschewed collections, these 'science centres' are important to the story of science museums in the twentieth and twenty-first centuries, so let us slow our historical pace for a while to consider one of them in particular.

The brothers Robert and Frank Oppenheimer cast long shadows over physics in the post-war United States. Robert was, famously, scientific director of the Manhattan Project and known as the 'Father of the Atomic Bomb': even if that may be an overstatement given the collaborative nature of this giant project, then certainly he was a favourite uncle. Beside him at the 'Trinity' atomic test was his younger brother Frank, who was of a more hands-on bent and gravitated towards experiments rather than theory.[40] Unfortunately for Frank's career in science, he also dabbled in communism with his wife Jackie, which came back to haunt them during the McCarthyist purges in the 1950s. He was called before the House Un-American Activities Committee and compelled to leave his university post (Robert too was later stripped of his security clearance). They retreated to Colorado, where Frank eventually found work at Pagosa Springs High School,

28 New kid on the block: the Exploratorium early in its existence, probably in 1969.

where he built a suite of improvised and borrowed apparatus. He then elaborated upon them when he moved to his next post at the local university (complete with his FBI tail). There he established a 'library of experiments' (ironically, given that it was intended to replace book learning). In 1965 the Oppenheimers visited the holy trinity of the Deutsches Museum, the Palais de la découverte and the Science Museum. Frank was inspired by their hands-on spaces, but decided he could do better.

Moving to San Francisco, Frank and Jackie planned a museum-that-was-not-a-museum in the building that had been erected as the Palace of Fine Arts for a 1915 expo, secured for $1 a year with his oft-used charm. With a handful of grants, largely improvised equipment and a young staff, the 'Exploratorium' opened quietly in 1969 (illus. 28); a TV programme later dubbed it the 'Palace of Delights' (redolent of a brothel, thought Frank). Frank's cousin, the philosopher Hilde Hein, spent a summer at the Exploratorium:

'visitors are not told how the world is,' she wrote, 'they are invited to find out for themselves.'[41] Young 'explainers' talked to visitors around stand-alone exhibits, each designed for the users to establish a specific principle for themselves – to facilitate questions rather than provide answers. In 'Eyeballs' (illus. 29), for example, the visitor is prompted to find out about binocular depth perception.

As Oppenheimer intended, the Exploratorium was indeed a departure from, and complement to, school science.[42] But it was not as freestyle and unstructured as perhaps he and his successors implied when they proclaimed that 'no-one flunks a museum'. The anti-expert approach required great expertise, and of course the explainers to explain it; the anti-collections approach required a large collection of apparatus. There can be no doubting the influence of the Exploratorium; it has spurred many imitators and has distributed the design for exhibits around the world for free. Now located on the Waterfront in San Francisco, it is a massive operation with a budget in the tens of millions. It was not, however, the first 'science centre' and Frank Oppenheimer did not invent interactivity, as we have seen above. The Exploratorium itself was not objectless, acquiring historical equipment from early on first by loan and then on a permanent basis, including a model early aircraft, a section of a linear particle accelerator, and an assortment of historic carpentry tools; in short, classic elements of a science and technology *collection*.

Oppenheimer was not working in a vacuum in the 1960s. His was only one of a raft of new or redeveloped institutions symptomatic of attitudes to leisure and science education in Europe and North America in the 1960s and a revivified vogue for international expositions. Up the coast, the Pacific Science Center began life as the Science Pavilion of the 1962 Seattle World's Fair; similarly, the New York Hall of Science began life two years later as part of the World's Fair. (It was the director of the former, zoologist Dixy Lee Ray, later governor of Washington State, who had coined the term 'science center' in the 1950s.[43]) The Science Museum relaunched its Children's Gallery in 1969, and a member of staff there was involved in another significant development in

29 Interactivity was core to the Exploratorium from the outset: here in the 'Eyeball' exhibit, illustration by Jad King.

North America. William O'Dea had pioneered visitor-orientated design principles and working exhibits in a number of large galleries in South Kensington – including an innovative electrical illumination exhibit as early as 1936 – and then applied them at the new Ontario Science Centre, Toronto, which opened as the Centennial Centre of Science and Technology in 1969. Its emphasis on participation and the dissemination of experiment design was as influential as the Exploratorium in many ways. Notably, in its early years it had a larger object collection than its young peer in Ottawa, the new National Museum of Science and Technology, which also had strong hands-on elements from its founding in 1967.[44]

Whichever exemplar was more important, there can be no denying the expansion and popularity of this approach to exhibiting science and technology over the following decades. The ferment of the 1960s quickly gave rise to new growth. Twenty North American science centres were sufficient critical mass to form an association as early as 1973 (only followed by their European peers in 1990).[45] Their combined visits more than doubled in the following two years, and this approach spread across the Atlantic. One channel for this was the Edinburgh-based perception psychologist Richard Gregory, whose ideas informed the vision exhibits that were central to the Exploratorium. Gregory himself had always been a tinkerer, as a neighbour remembered, and after he moved to the University of Bristol in 1970 he began to apply his combined skills to establishing a British equivalent of the Exploratorium.[46] With roots in the late 1970s, he formally established the 'Exploratory' in 1984 (moving into a railway goods shed in 1989, later becoming 'At-Bristol'; at the time of writing it is known as 'We the Curious').[47] Arguably the first science centre in the UK, it was followed by Techniquest in Cardiff and Satrosphere in Aberdeen (named for the Science and Technology Regional Organisation, SATRO). By the mid-1990s there were thirty British science centres. Like the Exploratorium, the Exploratory and its ilk used younger people as 'explainers' to facilitate stand-alone 'Plores'. This growth was mirrored in mainland Europe, for example in the founding in 1986 of the vast (and expensive) Cité des sciences et de l'industrie in Paris in the shell of an unfinished abattoir, the largest science centre in Europe.

The Cité des sciences is now run alongside the veteran Palais de la découverte as 'Universcience'. Amidst this flurry of new institutions, it is easy to overlook that among the early manifestations of what we would recognize as science centres were dedicated spaces within established museums. Among the earliest in Britain was a new interactive human biology gallery in the Natural History Museum. 'Launchpad' opened in 1986 as the successor to the Children's Gallery in the Science Museum, having first been mooted in 1978 and gathering steam after staff visited their former colleague William O'Dea at the Ontario Science Centre in 1981. It was intended to be 'a quantum leap forward in the idea of visitor participation'.[48] The short-lived 'Technology Testbed' in the Merseyside County Museum opened in the same year, 'Xperiment!' at the Museum of Science and Industry in Manchester in 1988, and 'Light on Science' at the Birmingham Museum of Science and Industry, opened by maths champion Carol Vorderman in 1988. The Royal Scottish Museum had a Discovery Centre for its young visitors in this period; although it proved 'equally appealing to adults and children, casual visitors and school classes'.[49] In America, too, the Association of Science and Discovery Centers included veterans such as the Franklin Institute and the Chicago and Boston science museums, pioneers of hands-on exhibits. These had all the hallmarks of science centres: they were designed to help their young visitors to ask questions, to find out how science *works*. But they were embedded in traditional museums. While its proponents were distinguishing themselves from collection-based institutions, then, many of the most prominent examples of the science centre movement were to be found as cuckoos in the nests within them.[50] Science collections continued to serve their function.

Historical Collections

While a new generation of gallery spaces emerged in the 1960s, a new generation of professionals were putting science collections to historical uses behind the scenes in tandem with developments outlined above. Grand new buildings tell one history of museums: to understand the functions of collections we need to look at how they were used, not only in display but in scholarship. And here we see clearly the influence of the passion of the curator on collections. The history of material culture is a human story.

To find the roots of the quiet heritage renaissance we need to go back two decades to a refreshed interest in history after the Second World War. Maurice Daumas, who would go on to be chief curator of the Museé des arts et métiers, wrote the first major analysis of the history of scientific instruments, and he introduced academic rigour to its exhibits and research.[51] At the same time in

Oxford another historian of science, Frank Sherwood Taylor, took up the post of curator at the Museum of the History of Science. He had been a chemist, inspired by his experience of being wounded at Passchendaele in the Great War. He then went on to be director of the Science Museum (albeit unhappily), where among his appointments he trained a young Frank Greenaway, who as keeper of chemistry would carry on this historical emphasis in his own writing and exhibitions.[52]

Greenaway was one of a generation of historian-curators who worked within and alongside an increasing number of university-based historians of science as the humanities expanded in the 1960s. The Deutsches Museum, at the forefront as ever, set up joint history of science and history of technology institutes with universities. Staff at the Science Museum wrote up doctoral theses on abstruse historical topics. Many historic instruments were displayed or redisplayed in new, heritage-focused galleries such as the chemistry gallery at the Science Museum in 1964, refreshed in the 1970s. In Washington, DC, trained historians of technology joined the self-taught curators on the staff of what was now the *National* Museum of History and Technology. They included renowned bibliophile, horologist and instrument scholar Silvio Bedini, as deputy director and, on several occasions, acting director. New recruits in the late 1960s included historian of engineering Otto Mayr (who would later lead the Deutsches Museum). During his time as director, engineering historian Robert Multhauf edited the history of science journal *Isis* and insisted his staff publish scholarly historical work. With good cause, he considered the museum to be the principal centre for history of technology in the United States.[53]

Towards the end of the 1970s the cadre of historically trained curators included a number of women, including Deborah Jean Warner in Washington (illus. 30) and Alison Morrison-Low in Edinburgh. They were part of a new wave of specialist curators and other museum professionals across the collection spectrum. Morrison-Low was instrumental in establishing a new gallery of historical apparatus in 1986, richly packed with 2,000 instruments

30 Smithsonian curator Deborah Warner with an armillary sphere in 1978.

from a twelfth-century astrolabe (illus. 31) to a Victorian petro-logical microscope.[54] By this time, at the helm of the newly formed National Museums of Scotland was Robert Anderson, who had started his career as a junior science curator in one of the institu-tions it incorporated, the Royal Scottish Museum. He had spent a decade in the Science Museum in the interim and worked with Greenaway on the chemistry galleries there. Anderson remained an enthusiastic historian of science and combined his directorial responsibilities with chairing the international Scientific Instruments Commission.

Such arcane endeavours could even be found in the newer institutions: the Cité des sciences in Paris included a centre for history of science research.[55] Similar developments were evident at the Science Museum, on which we will now narrow our focus for clarity as we approach the busy-ness of the present. There, history of science endured under the new umbrella National

Museum of Science and Industry (led by former Ironbridge director Neil Cossons) alongside industrial heritage on the one hand and interactive experiences like 'Launchpad' on the other. Cossons oversaw a rigorous restructuring in the late 1980s, which resulted in a new Research and Information Services Division, headed initially by Robert Fox, a respected academic historian of science who had directed historical research at the Cité des sciences. Fox's brief was to stimulate academic publications: but the new(ish) generation of historians of science in museums expressed their

31 A brass astrolabe collected and displayed in Edinburgh as history re-established a firm footing in science collections. This is the earliest surviving signed and dated astrolabe made in Europe, by Muhammad ibn al-Saffâr, Cordova, Spain, early 12th century with 13th-century rete (the pierced front plate).

32 Big scientific kits 'rescued' by curators included the Cockcroft–Walton generator in the National Museum of Scotland – here with eager visitors in 1989.

interests not only in publications and exhibitions but, of course, by collecting.

The trade in antique instruments gathered pace, and curators also undertook 'rescue' collecting of recently obsolescent scientific equipment (illus. 32), from mainframe computers to nuclear power station control panels. But whereas auction rooms and laboratory skips fuelled science collections considerably, the largest act of collecting of any science collection at any point overshadowed them both. Pharmaceutical magnate Henry Wellcome had gathered the largest private collection earlier in the twentieth century, buying up medical history as well as voraciously (and uncritically) amassing archaeology and ethnology. In terms of wealth, clout and collecting impulse, Wellcome was only rivalled in twentieth-century science, technology and medicine by industrialist Henry Ford, who gathered a huge collection of American technology as a giant 'animated textbook'.[56] Whereas Ford's collection remains in the museum that now bears his name, it took three decades after Wellcome's death in 1936 to select, sort and debate the fate of his collection, already spread over several museums and warehouses. (It was eventually dispersed between 1978 and 1983 in the only de-accessioning process I am aware of that was measured by the tonne.) Frank Greenaway, interested in the history of medicine, thought the Science Museum might be a rather good home for some of it. And even though only a fraction of the collection remained, the very specifically medical element still involved more than 100,000 objects. When the Science Museum accepted it – on 'permanent loan' from the Wellcome Trust – it doubled the quantity of objects in its collection. Two large galleries would eventually result. The shape of this great collection is evidence of the whims of the collector and the interest of the curators who accepted it.

Understanding Collections

It has been clear from this chapter that historical and contemporary currents have flowed through science collections since their inception. Tensions between reified relics and hands-on interactives swung back and forth during the twentieth century; it is, then, no surprise that debates around the functions of science museums (and specifically how far they should be about the past or the present) continued at the end of the century. The final swing of debate in the twentieth century was an approach known as the public understanding of science (a renewed emphasis on a term that Oppenheimer had used).

In the early 1980s members of that august association of the greatest British scientists, the Royal Society of London, were concerned that the general population did not understand (much less appreciate) their great work. So they did what learned societies do best: they convened a committee. Chaired by geneticist Walter Bodmer, the esteemed group reported in 1985 and implored scientists to communicate with the public via television and other popular media. In the consumer-led 1980s they linked scientific education, informal and formal, with national prosperity and extolled the virtues of informed decision-making.[57] Although the group included Science Museum director Margaret Weston (as well as David Attenborough, who was very familiar with natural history collections), Bodmer's report barely referred to museums. He nodded to science centres, but only mentioned the Exploratory by name. The report, however, also gave rise to the Committee on the Public Understanding of Science (COPUS), involving the Royal Society, the Royal Institution and the British Association for the Advancement of Science. COPUS, which was wound up in 2002, in turn set up a working group to encourage science centres.[58]

The main proponents of PUS (an unfortunate acronym that stuck) may not have paid very much attention to museums, but museums paid attention to PUS. They had, after all, been communicating to the public about science for some time. Since the Great Exhibition and before, science collections were deployed for the citizenry to better understand the natural world and the industries that exploited it. Museums in the 1980s were keen to be part of the latest vogue: the Cité des sciences, for example, identified itself within this framework, as did the Chicago Museum of Science and Industry.[59]

In the UK, the Science Museum was especially enthusiastic. Robert Fox's successor in the research division, John Durant – himself a member of COPUS, later director of the MIT Museum – took up the crusade in earnest. During his tenure his domain was retitled the 'Science Communication Division'; he edited a new journal, *Public Understanding of Science*; and initiatives like 'Science Box' temporary exhibitions presented the latest science (separately from the permanent displays).[60] Curators and historians wanted to use collections and the stories to be told with them to *humanize* science, as a process, an activity like other human activities: to 'open up the house of science'.[61]

This momentum carried forward into the 1990s. The new European group for science centres and museums known as Ecsite (European Collaborative for Science, Industry and Technology Exhibitions), with Richard Gregory at its helm, explicitly positioned itself as a network in this area.[62] The Museé des arts et métiers was redeveloped in the 1990s in the PUS vein, and in South Kensington the new Wellcome Wing had an explicit focus on the contemporary, especially around biomedicine and information technology. In the United States, likewise, concern about national science illiteracy was heading towards a perceived crisis, and science museums increasingly took a visitor-orientated approach to exhibitions and programmes, now framed within 'informal science education'.[63] On the Mall, the Museum of History and Technology was now the National Museum of American History, but a new Center for the Study of Invention and Innovation in 1995 blended its traditional strength in the history of technology with efforts to inspire creativity, later via the 'Spark!Lab'. Patron Jerome Lemelson wanted to make innovation 'exciting to young people'.[64]

Science centres had surfed the PUS wave through the 1990s and were politically expedient; several were explicitly part of urban regeneration schemes, including, for example, the Maryland Science Center in Baltimore. China in particular made massive investment in new science museum-cum-centres based on the Exploratorium model at the beginning of this century, aiming to establish two hundred new institutions.[65] Of more than five hundred worldwide, in the UK by this time there were some two dozen stand-alone science centres, of which half were established at the millennium thanks to a £250 million boost, described by Gillian Thomas, the chief executive of At-Bristol, as 'the largest single investment in science communication ever to take place in this country'.[66]

And yet, as the UK's House of Lords committee on the subject admitted, there was little evidence that any of this had made an impact in the understanding of science (however that might be defined) in the general public (however they may be defined); measuring its success was, without irony, 'an inexact science'.[67] Their Lordships called for a change from the 'public understanding of science' to 'science engagement'; and, indeed, this was the direction of travel in the sector in the years since.

Hybrid Collections

In the words of one participant, the PUS movement 'came to a shuddering halt at the end of the 1990s'.[68] Not all the new centres survived the fading of the millennium buzz. Meanwhile, science museums with collections – many with their own internal science centres – stubbornly endured. They faced challenges, of course, but their long dual-purpose history served them well, and they survived to fulfil the functions we will explore in the chapters that follow. And so, like Foucault's pendulum, science museums throughout their history swung back and forth between past and present, between handling and looking, between science and industry. Before exploring where the pendulum is now, we might pause to reflect on what this historical sojourn has revealed about science collections: that they are idiosyncratic; that they are political; and that they are hybrid.

Science museums may at first appear to be universal, promoting and explaining eternal truths, regardless of place or personality. But if this trip through their past has revealed anything, it is that their development is highly serendipitous. They rely especially on the particular passions and positions of their founders, and we have met some peculiar folk along the way: from the eccentric Henri Grégoire, to the dogged civil servant Henry Cole, through the mover-and-shaker Oskar von Miller, to incorrigible physicist Frank Oppenheimer, founders' eccentricities are embedded in their institutions. Their choices of location, their choices of things, will shape the collection: for as curators are fond of repeating, they house what was selected to survive rather than what there was. For even though their successors may have spent more time, added more things, built larger buildings, the venerated Founding Fathers (there are precious few Mothers) will cast long shadows – and their myths are perpetuated by those that follow.

In exploring the human element of science museums, we have found how much politics shapes these supposedly neutral entities. This we can see in their founding: the Victorian culture of improvement and the South Kensington Museum, early twentieth-century nation- and empire-building and the Deutsches Museum, and the swinging sixties that gave rise to the Exploratorium. And they have continued to be subject to nationalism and economic currents during their subsequent existence. They were tools of the Cold War, for example, especially in promoting the space race. During the 1980s UK museums explicitly reflected political mores as they charged entry for the first time, framing the visitor not as a student but rather a customer. As Margaret Thatcher proclaimed of British ingenuity, past and present: 'I am delighted that the Science Museum also displays the most modern techniques – for example in the new galleries on the Chemical Industry, Plastics and Space Technology.'[69] We will return to the myth of neutrality (and another divisive British prime minister) in the final chapter.

Perhaps it is no surprise that these otherworldly places are as human as any other institution. They are shaped by individual idiosyncrasies, serendipity and happenstance; they are contingent and locally determined. And they are also riven with tensions as to their purpose. Throughout their history, as we have seen over and over again, they have been pulled back and forth – to be about past and present, to be for scholars or schoolkids, to be austere sanctuaries celebrating achievements or hands-on playpens. Anthropologist Sharon Macdonald has written of the troubled relationship between 'the yawn-provoking Scylla of traditional "brass and glass" presentation' on the one hand and the 'possibly vacuous "whizz-bang" Charybdis of the objectless science centre' on the other, but this was not new.[70] The tension between historical museums and science centres was over-egged: science collections have always been part heritage, part hands-on. Foucault's pendulum continues to swing.

33 OncoMice, or Harvard Mice. Freeze-dried male transgenic mice 1134 and 1136, direct descendants of the first mammals to be granted a u.s. patent, #4,736,866, dated 12 April 1988. They are now in the Science Museum in London.

2

COLLECTING SCIENCE

T hese two freeze-dried male mice (illus. 33) are not, as one might expect, in a natural history collection, but instead they are on display in London in the Science Museum's 'Making of the Modern World' gallery. For these are no ordinary rodents. They are Harvard Mice, more famous as 'OncoMice', the first kind of animal to have genetic material introduced from another organism. Sadly for them, this transgene made them much more susceptible to cancer; happily for us humans, the Harvard laboratory then used them to understand the disease. As transgenic organisms 1134 and 1136 they were important tools for clinical research; as descendants of the original mice involved in United States Patent 4,736,866 in 1988 (the first mammal to be granted a patent), they were part of a complex system of intellectual property; and as Science Museum object number 1989-437, acquired in 1989, they are important museum specimens, boundary objects spanning organism and artefact.[1] Why did the Science Museum collect these particular object-organisms of the many that could have been? And what can they tell us about science museums?

So far in this book we have looked at the 'what' and 'when' of science collections: here we will consider 'how'. This is not a simple matter. Collecting scientific, technical and medical objects can be difficult, expensive and often thankless. And one might assume that of all museum disciplines, science is collected, well, scientifically: surely scientific things are gathered dispassionately and comprehensively? As one former curator observed, 'Visitors tend to believe that we collect objectively, eternally.'[2] Surely we must have moved on from the quirks and idiosyncrasies we found in the last chapter? Surely science museums in the twenty-first century are no longer subject to individual personalities and bias?

It will come as little surprise that science collections, like any other, rely on particular interests, local forces and serendipity. By unpacking the everyday collecting practices, we can better understand how collections are shaped, and the role of curators in selecting the objects that will remain in them in perpetuity. We find that collections continue to be subject to the tensions we have already encountered: historic relics jostling with contemporary tools; the vast and the minuscule side by side; the remains of great individual experiments versus examples of everyday scientific practices. Exploring the challenges and benefits of these tensions will show that collecting is fundamentally human and partial as an activity, and the results are profoundly human and partial – but all the better for it. Museums collect old and new, tangible and intangible, but most of all, they collect stories.

The OncoMice are among the flock of objects we will stalk through the trials and tribulations of collecting science, from an exquisite seventeenth-century clock to material related to the COVID-19 pandemic. Their afterlives will illuminate first how curators like Robert Bud of the Science Museum or Ingenium's David Pantalony select objects and the challenges they face; then the different ways objects arrive, whether via 'fieldwork', given, purchased or through digital channels. As well as collecting techniques bringing things into collections, we will also see that this flow is sometimes reversed, that science collections shrink as well as grow. Once again, we find science collections are not the static mausolea they may appear, but rather they are lively, active entities.

What to Collect

Science collections develop unevenly, growing faster at times of formation and at the peak of imperial collecting. Whereas their predecessors had ambitions to collect encyclopaedically, seeking representative examples of everything in the natural and cultural worlds, the twenty-first-century curator is more selective. Collecting happens in fits and starts, according to project timescales, new buildings and the like. For example, during much of the period of preparing this book, the Canada Science and Technology Museum had imposed an acquisition moratorium while they built their new Ingenium Centre (a shiny new facility we shall discuss in the following chapter). Objects nevertheless do continue to arrive in science collections around the world – from mice to microscopes, from vacuum cleaners to voltmeters.

A moderate but active specialist scientific instrument collection like the Whipple Museum in Cambridge tends to bring in up to one hundred items per annum. A large collection like the Science Museum or the Deutsches Museum will absorb several hundred physical and digital items each year, as well as contextual material including interviews, photographs, documents and manuals. To put these in a comparative context I took a snapshot of these three institutions' collecting activity over a recent year and compared it to the relevant department in my own museum (illus. 34).[3] A year's acquisitions still represents only a fraction of 1 per cent of the total collection size of any of them, which, as we saw in the introduction, runs into the hundreds of thousands. To put the annual growth of science collections into context, these figures overshadow the dozens of works that might enter a fine-art collection, but would be dwarfed by the tens of thousands that a natural history museum would collect.

The nature of the new arrivals will span the spectrum of material we first encountered in the introduction. Some are extraordinary objects like an OncoMouse, which the Science Museum classes as an 'icon', or associated with what the u.s. National Museum of American History considers representing 'seminal events', whether social or scientific.[4] Some are associated with local heroes, be they from the country, region or nation represented; others are of international importance. Ingenium in Canada terms these 'benchmarks', for example Canada's first laser, which curator David Pantalony acquired in 2010 (illus. 35). Physicists Boris Stoicheff and Alex Szabo had crafted this ruby crystal surrounded by a glass tube in 1960 and switched it on only briefly in 1961. It then spent a decade in a drawer followed by forty years on display in the laboratory at the National Research Council. It was a benchmark even before it came into the museum.

But for every world first there will be many more from the ordinary everyday. These reflect the practice of science, the use of technology and transport, and the experience of medicine. Humble items recently collected by the Science Museum include a teacup for the medical galleries that demonstrated the way polio vaccination practice was disseminated in India. The

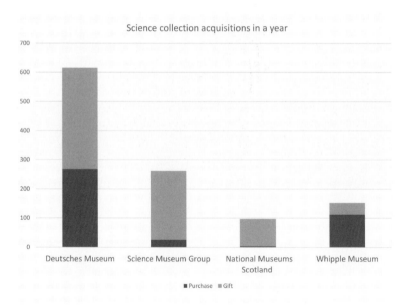

34 A sample of the quantity of acquisitions from four science collections; the Deutsches Museum in 2017 and the Science Museum Group, National Museums Scotland (Science and Technology only) and the University of Cambridge Whipple Museum of the History of Science in 2018. Quantities are not always the best measure, but these give an indication of the scale of collecting and the balance between gift and purchase.

35 The first functioning laser in Canada in 1961: a rod-shaped ruby crystal surrounded by a fast-discharge flashtube, held in an aluminium cylinder; now in the collection of the Canada Science and Technology Museum.

majority of the items that comprise the new acquisitions (see illus. 34) will tend to be small, and to have arrived in sets. Among the Science Museum Group's acquisitions in the sampled year were twenty historic luggage labels and 75 railways coats of arms, for example, while National Museums Scotland acquired a large batch of electric measurement apparatus from Scottish Hydro Electric. Likewise university museums acquire in batches:

36 A 16th-century tide computer attributed to the scientific instrument maker Charles Whitwell, purchased by the Whipple Museum.

local Nobel Prize-winners' apparatus went to the collection 'Gustavium' at Uppsala University in Sweden, for example, or in Canada nineteenth-century meters arrived at the historical collection when Queen's University's physics department moved premises.[5]

Some are rare, some are common; and like existing collections, some items are old, and some are new. The ancient-versus-modern duality that ran through their histories in the last chapter remains evident today. In the snapshot survey I found that 7 per cent of the objects collected by the Science Museums Group were of twenty-first-century origin, and 17 per cent for the Deutsches Museum. Among new acquisitions in Edinburgh in that time was a batch of humble ceramic nuts and bolts, straight from the production line, for a new exhibition in which they evidence the versatility of ceramics today; the Whipple Museum in Cambridge, by contrast, was busy acquiring from a scientific instrument dealer an exquisite Elizabethan device to calculate the time of the tides and to convert solar, lunar and Zodiacal calendars (illus. 36).[6]

What links these objects, be they old or new, rare or common, is their significance, scientific or otherwise. Significance is not absolute: different elements are foregrounded in different sorts of collections – art, history, natural history – and what is considered significant changes over time. Whereas a nineteenth-century technology museum might have looked for examples of national innovation, for example, in recent years, science museums collect to illuminate the relationship between science and society. Artefacts' significance incorporates social, cultural and aesthetic as well as technical factors, and there are of course many non-scientific elements at play in the selection of items. Manchester's Science and Industry Museum, for example, represents 'the ongoing interplay between science, industry and society', and likewise, Ingenium collects 'the products and processes of science and technology and their economic, social and cultural relationships with society' and will focus on 'the relationships between people and science and technology'.[7] These artefacts are all intended to connect science to (the rest of) culture. Unsurprisingly there is a strong human element in judging these connections. Ingenium curator David Pantalony observes, 'Collection work is not all rational . . . we must trust the non-intellectual dimensions of being out in the field. What is surprising and fascinating to the informed curator in the field, may have great value for the collection and museum exhibits.'[8]

With these enduring tensions in mind, let us now peer behind the curtain and see how curators do what they like to do best: collect things.

Challenging Collecting

The curator faces considerable challenges in collecting scientific things. Before moving on to reveal the modes by which they acquire things if they succeed, we will consider the difficulties presented by science objects' size and complexity; but first, their bewildering quantity.

The iPhone is a seemingly omnipresent technology in the twenty-first century (illus. 37). It is clear that at least one belongs in a museum, but which one? The abundance of material culture since the Industrial Revolution confounds the curator. A recent research project addressed this very problem, 'profusion': 'How in the face of there being so many more things produced today – beginning with industrialisation and mass-production, especially since the mid-nineteenth century, and then accelerated by post-Fordist production since the 1970s – is it decided what will be kept for the future?'[9] There is an abundance of things available already, and also new products arriving rapidly on the market. Technical artefacts can be commonplace, but had former curators not collected them, there would not be a record of the historic practice of science. If curators don't follow their lead today, there will be a generation missing from the historical record.

One solution to this problem of profusion is to select early and select boldly. Take, for example, that most endemic tech company, Apple. Of more than one hundred Apple products in the National Museum of American History, perhaps the most impactful is the simple adapter found at the site of the World Trade Center after the attacks of 11 September 2001. At National Museums Scotland we have been collecting iPhones for a decade, a nice extension to our existing communications collections, from telegraphy onwards. But from the two dozen models and the billion handsets sold, which would we collect? We went for those that had stories. One was a prize in an online competition, arriving just before the first iPhone launch on the European market, which must make it a contender to be one of the earliest examples in the country. Another was used by a prominent videogame designer, Mike Dailly of YoYo Games, to develop *Simply Solitaire*, which at its peak was Apple's most popular free title. A third belonged to photojournalist David Guttenfelder, which he used to take award-winning images of conflict zones for *National Geographic*, other agencies and online channels such as the

37 A first-generation Apple iPhone and its original packaging acquired from the USA as a prize in an online competition and shipped to Scotland in November 2007 prior to European release.

photo-sharing platform Instagram, of which he was an early leader. And so although one might think that a handset might be dull for a museum (perhaps the same might be said at first glance of the OncoMouse?), all phones are unique once they are in the hands of their users. The stories that go with them help us to select; they provide a route out of the maze of profusion.

iPhones may be difficult to choose from, but they have the advantage of portability. Many elements of science and technology are not so handy: steam engines, whether stationary or locomotive; early computers the size of buildings; industrial

38 The prototype suspension system for Advanced LIGO gravitational wave detector with dummy metal test masses collected by the Science Museum.

equipment; particle accelerators. The Imperial War Museum is not the only museum organization to have an object that is also a site: in their case HMS *Belfast*. They also tried to acquire a nuclear submarine in the 1970s.[10] Science collections have many elegant examples of reflecting and refracting telescopes, but fewer of those that stretch over tens of metres (which go by the cunning initialism 'ELT' – extremely large telescopes). One of the most important scientific discoveries of recent years, the Nobel Prize-winning detection of gravitational waves, uses a pair of instruments that have laser beams projected down 4-kilometre-long (2½ mi.) 'arms'. These are tricky to collect, so the Science Museum borrowed a prototype beam-splitter, which at least fits in a room (illus. 38). Something so brutally pragmatic as space determines much collecting; and museology, like nature, abhors a vacuum, so museum stores have a habit of filling up.

The curator must be cunning in representing the massed materiality of technology. When the Science Museum collected a chunk of the Large Hadron Collider, curators also acquired one of the bicycles used to travel around this vast apparatus, neatly encapsulating the human scale and element of the scientific endeavour.[11] One of my favourite artefacts in any collection is at National Museums Scotland: the flare tip from the very top of the Murchison oil platform. When oil is extracted, raw natural gas comes to the surface as well and is burnt off as unusable or waste gas, and this is the flare used to burn this off at a safe distance from the main rig. Upon decommissioning, curator Elsa Cox persuaded the oil company to save it from (lucrative) scrappage (illus. 39). Only a tiny fraction of the total artefact, it is nonetheless impressive close up, twice my height and weighing 800 kilograms (1,765 lb). One would be hard pushed to find a more powerful way of illustrating the sheer scale of the North Sea oil industry and its decommissioning, and all that meant for the UK. It is a part that represents the whole, a synecdoche, which is a handy tactic in museums with limited space. Again, it is the story that we collect, by gathering video and artworks and testimonies of life on the oil platform.[12]

39 Curators and bemused salvage professionals consider the flare tip from the Murchison oil platform from the northern North Sea, 2017. It is between states: disassociated from its operational maritime context but yet to become a museum object. The image shows suggestions as to how to dismantle a manageable portion.

Sometimes the size-related problem is not massive-ness, but the opposite. In recent years nanotechnology – manipulating matter on the scale of billionths of a metre – has become increasingly important. Sarah Baines at the Science and Industry Museum in Manchester collected around graphene, a substance that had been identified in the nearby university in 2004 and which comprises a single layer of carbon in a strong hexagonal lattice. In 'Wonder Materials' she and her colleagues displayed the sticky-tape dispenser used by one of its developers, Konstantin Novoselov, to pick up this invisible substance (illus. 40).[13]

Small and large, technical artefacts are often rather complicated, difficult for curator and visitor alike to grasp. Many scientific instruments suffer from the 'black-box' syndrome; they comprise incomprehensible systems, often electronic, in featureless cases. David Pantalony wrestled with this when he collected a surgical operation recorder (which was actually a blue box). Developed in a Toronto hospital, the device captured sound and vision, by parity with a flight recorder on an aircraft. David collected it to represent this 'bold and novel practice that takes aim at traditional notions of surgical culture, tacit expertise and authority', but was conscious that it was 'remote, abstract, mysterious, boring, and not human'.[14] Its parts were made all over the world (illustrating the internationalism of science and technology), and some by companies with military interests, thus highlighting the complex systems of patents, intellectual property and secrecy that scientific objects bring with them. (Another museum displayed the empty shell of a pacemaker because the company did not yet want to donate the commercially sensitive workings.) And, crucially, David was not immediately able to collect the surgical blue box because it is still in use: rather, he tagged it to come back for later. Curators appreciate that science is an unfinished business.

40 Graphene, the wonder material: the 2002 tape dispenser used to pick up the tiny substance, displayed in Manchester in the 'Wonder Materials' exhibition in 2016 at the Science and Industry Museum.

How to Collect

Undeterred by profusion, size and complexity, in 2013 David made the 4,000-kilometre round trip from Ottawa to Winnipeg. Like curators of other collections, he was undertaking fieldwork: neither scraping for fossils nor catching mudflies, but rather engaging with scientists on their home turf to snare their kit. This is one of the different routes by which he and other curators obtain new material, which may be gathered, bought, given, borrowed or downloaded. Let us follow some objects' paths to the collection to reveal the tricks of the curator's trade. And, of course, their colleagues' tricks too; in museums lucky enough to have specialist staff, curators work with registrars to process the accessions, conservators to assess and stabilize them, and collection managers to store them.

David was in Winnipeg to visit University of Manitoba physicists, especially those researching time-of-flight (TOF) mass spectrometry, and he had his eye on their apparatus for innovative analysis of biological molecules. 'Some scientific facilities carry both deep heritage elements, as well as a compelling sensory experience that is worthy of noting and preserving,' he later reflected.[15] He 'surveyed a vast landscape of electronic equipment' and on the advice of the scientists he selected the 'TOF2', a 3-metre-high (10 ft) mass spectrometer from around 1990. He also earmarked for later acquisition, once its use-life concluded, a section of the Manitoba II: a bespoke, room-sized instrument on which the team had set international standards for atomic masses (illus. 41). 'The room and instrument document over forty years of toil and triumph,' David observed. 'There are shelves of log books, abandoned parts, tools, signs, layers of black board sessions, trade literature, texts and aged off-prints. The instrument shows countless modifications, inscriptions, warnings, heat streaks, and tape – lots of tape.'[16] Both this supporting material and the improvised character of the machine were appealing, providing the context and, crucially, the human stories associated with the equipment. Curators like David become expert brokers, cultivating these

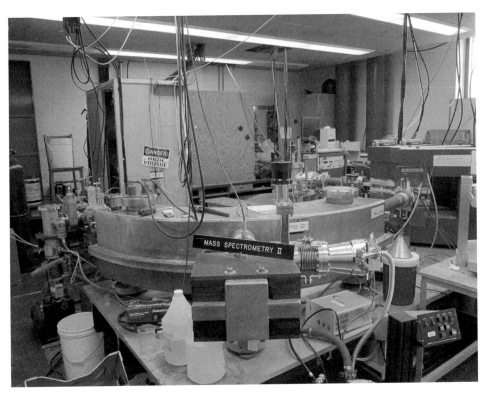

41 Fieldwork: the Manitoba II mass spectrometer *in situ* in the University of Manitoba Physics Department. Curator David Pantalony has his eye on this for the Canada Science and Technology Museum.

relationships and knowing whom to ask for what, and when, in order to extract these tales.

David's visit was part of a wider concerted effort, but sometimes collecting fieldwork is in response to an emergency call from an institution telling a curator they have found something potentially of interest, but that there is a limited time envelope before the bulldozers come in. National Museums Scotland curators have a knack of swooping into decommissioned power stations in the nick of time.[17] Admittedly they walk off with very little from the mass of machinery, but one can tell powerful stories with something as simple as the telephone that linked the national grid to Longannet power station, which once stood on the shore of the Firth of Forth (illus. 42). Other times there is less urgency: curators were able to visit CERN at a more careful pace

42 A telephone used to connect the Longannet power station to the National Grid, collected on the eve of demolition for National Museums Scotland.

to consider collecting the copper cavity we met at the opening of this book (illus. 43). Roaming curators also find interesting old instruments in out-of-the-way places. As Dartmouth College's Richard Kremer reflects of the collections in his university:

> Squirreled away, the retired items sit silently, collecting dust on back shelves or in storage cabinets near the laboratory spaces where they formerly enjoyed more active lives. Occasionally janitors, lab technicians, retired professors or other hoarders aggregate the obsolete equipment into specially designated spaces, invariably those basement rooms of little value . . . Hoarded objects remain liminal, precariously poised between active life in the laboratory and death in landfill or scrap metal recyclers.[18]

The canny science curator focuses on material that is old enough to be obsolete but not so old as to be collectable.

The best way is to be proactive about it and make friends with the scientists, designers and users of the kit one would want to collect. 'We need to be active,' advised the doyenne of Portuguese university curators Marta Lourenço, and 'go meet the "generators" where science is made – laboratories, and factories, universities – and "field collect", instead of passively waiting for the next donation to come'.[19] Robert Bud of the Science Museum made contact with the Harvard researchers developing the Onco-Mice through personal connections, inviting them to the Science Museum, and in dialogue selected the appropriate individual mice and ultimately secured the gift. A highlight of my time in Scotland has been a visit to the Centre for Science at Extreme Conditions at the University of Edinburgh to see their Mars simulator, which we had our eye on. Even though it actually looks rather like a souped-up microwave, it was delightful and we

43 National Museums Scotland's Gordon Rintoul and Alexander Hayward with Peter Clarke (left), the physicist who brokered the museum's relationship with CERN that led to the acquisition of a copper cavity like this one. Also part of this collecting trip was Tacye Phillipson (behind the camera).

expressed our interest to collect it, if and when they no longer needed it. Like David Pantalony's approach with the surgical blue box, this is 'Post-it note' collecting; identifying interesting objects during their working life to be donated as soon as they are replaced.

We relied on the hospitality of astrobiologists when visiting the Mars simulator, just as physicists helped David, geneticists helped Robert and oil professionals helped Elsa. Collecting is a collaborative activity, especially fieldwork: scientists are science curators' source community. As well as doctors, patients, designers and tech users, they can be great sources of objects, but even better, they can provide the *stories* that go with them. Curators at the Science Museum worked with the International Astronomical Union to collect material representing the International Year of Astronomy, for example, and with Cameroonian phone salespeople for their communications gallery 'Information Age' (illus. 44).[20] It is important to remember that curators are only one part of this process.

If a dose of humility helps, so too does cold hard cash. In 1662 Alexander Bruce, 2nd Earl of Kincardine, designed a device that could keep such accurate time that it could determine longitude at sea (a crucial factor in rendering maritime travel safe and accurate) and commissioned Dutch instrument maker Severyn Oosterwijck to make it (illus. 45). The result was an adapted pendulum clock, the recent invention by Oosterwijck's countryman, mathematician and astronomer Christiaan Huygens. After sea trials (and a brush with piracy) the attempt was ultimately unsuccessful; the problem would not be solved for another century, by English clockmaker John Harrison. The Bruce–Oosterwijck clock then disappeared from the historical record for three hundred years, reappearing in the collection of a dealer who had considered it a bracket clock – exquisite, yes, but its original function forgotten.

Once its significance re-emerged, the clock, one of only two surviving, came on the market in the 2010s. After a couple of false starts, in 2018 National Museums Scotland curator Tacye Phillipson set out to purchase it from a private collector. She had help from

44 A mobile phone call box with advertising sign, used in Bamenda, Cameroon, 1997–2012, obtained for the 'Information Age' exhibition at the Science Museum in London.

45 An unusually extravagant purchase for National Museums Scotland: the Bruce–Oosterwijck longitude pendulum clock, part of the first attempt to establish longitude at sea with a purpose-made mechanical timepiece, the movement by Severyn Oosterwijck for Alexander Bruce of Kincardine, The Hague, 1662. The original mechanism – the expensive part – is in the middle.

museum registrars (who work hard on the legal and logistical elements) and conservators (who assess the physical condition of potential acquisitions), and even though funding is thin on the ground for science collecting, she secured support from the National Heritage Memorial Fund (a pot of money UK public museums can apply for to help purchase items for their collections) and even the Art Fund to help support the purchase, despite the clock's wilful plainness. It provides striking evidence of Scottish–Dutch collaboration, melding austere Calvinist aesthetic with the Netherlands' Golden Age. The clock joined the collection for a total price of £80,000: a sizeable sum, eight times the price of the tide computer purchased around the same time by the Whipple Museum (see illus. 36).

In another light the clock might be considered a bargain, given that at one point it had been valued at five times that. And even that inflated cost was dwarfed by prices at the upper end of the fine-art market. Multimillion-pound auction sales are news-worthy: the national galleries in London and Edinburgh purchased Titian's *Diana and Actaeon* and *Diana and Callisto* (both 1556–9) for a combined total of £95 million a decade ago. Buying at auction may be the highest-profile way of collecting but is rarer for science curators than it is for their art gallery colleagues: more often when they buy it is small purchases from dealers, makers, shops and even eBay. David Pantalony's colleagues at what is now Ingenium purchased on a more usual scale as part of their 'Memories Are Made in the Kitchen' initiative (illus. 46). Anna Adamek and other curators solicited suggestions relating to contemporary Canadian kitchen equipment via a crowd-sourcing call that went out to the museum's members, on social media and in the national press, prompting around 170 offers. Curators then selected eight items they considered most signif-icant and/or representative, including a particular brand of coffee maker and an electric wine opener. Buying them from high-street shops and online retailers, they documented the very act of purchase as part of the acquisition:

> All original packaging and trade literature, such as manu-als and catalogs that came with the appliances, were accessioned into the collection. The price of each appli-ance was recorded, and the purchase orders, invoices, and store receipts were kept in order to document the source of the appliance, the name and address of the store, the date and time of purchase, and the amount of taxes paid on the domestic kitchen equipment. Such documen-tation, it is hoped, will allow future researchers to gain a better understanding of the consumer's experience in relation to the displays and sales processes.[21]

The total came to c$1,075.92.

It is more common, however, for items to be given than bought. A steady stream of offers comes to museums, ranging from printing presses to defunct laptops. Statistically, most donated acquisitions are unsolicited, but occasionally there are relationships curators cultivate in the hope that they might bear fruit, or occasions when they actively set out to call for particular items. One such occasion was the collecting activity stimulated by the COVID-19 pandemic. In spring 2020 museums scrambled to close down, some for the first time since the Second World War. Most curators settled in to work at home, or prepared for life on 'furlough' (by which the UK national government supported organizations whose employees could not work). After a collective breath, many organizations saw the opportunity and importance of collecting the digital, visual and material culture of the impact of the virus, especially those with social history and medical collections. The National Museum of American History has both, and so formed a Rapid Response Collecting Task Force to 'document the scientific and medical events, as well as the effects and responses in the areas of business, work, politics and culture'. Like many museums, they were unable to accept objects while closed, but they asked would-be donors to retain material until the crisis passed. In the meantime, offers flooded in, from 'handwritten grocery lists and letters from patients to personal protective equipment, test kits and ventilators', as well as oral histories and other digital entities.[22] This mode of collecting – the public call – is not new but this one had a particular resonance and relevance. The curators were proactive and selective; many others installed Web portals to invite would-be donors to submit their ideas, object descriptions and memories.

While calls like this prompt gifts for particular projects from individuals, others arrive more quietly from organizations and groups. Both the OncoMice and TOF2 were gifts from universities to museums (Harvard to the Science Museum, University of Manitoba to Ingenium); so too the University of Edinburgh has on several occasions over their entwined histories transferred important material to National Museums Scotland and its predecessors. One founding element of the museum was the so-called 'Playfair Collection', given to the first director George Wilson by his friend Lyon Playfair, the professor of chemistry. The highlight of the collection is the glassware used by renowned chemist Joseph Black (illus. 47); but it is Playfair's name that has been indelibly associated with the collection ever since.[23]

Whether from institutions or individuals, whether considering offers or seeking to solicit gifts, it pays to think about *why* someone might offer an artefact to a museum, especially if the object is of considerable value. True, many are motivated by a genuine wish to do good, to preserve something interesting for posterity, to educate or entertain present and future museum audiences. In the case of the COVID-19 gifts, the urge to commemorate a shared experience is clearly very powerful. Or else the donor wants to establish a legacy, often not for themselves, but for a forebear (as Playfair wanted for Black) an institution or a profession.

46 Crowdsourced collecting: kitchen appliances acquired for the Canada Museum of Science and Technology.

47 Glass retort used by Joseph Black between 1766 and 1799, made by Archibald Geddes, Leith. Part of the 'Playfair Collection' transferred from the University of Edinburgh to the Industrial Museum of Scotland in the 1850s.

It is perhaps no secret that less altruistic motives may also be at play. Giving to a museum may simply be a way of clearing space in one's attic or laboratory. There may be a commercial imperative: what better way for a corporation to promote an invention or product than to display it for thousands of visitors? (And in any case, as we will find in the next chapter, donors are often disappointed to learn that the vast majority of things will not be exhibited.) To give to a museum is an act of appealing visibility, bringing prestige on the donor and connecting them to the object through the label and catalogue for evermore. As museologist Susan Pearce observed, 'to give material freely to museums is a meritorious act which conveys famous immortality ... the huge bulk of museum collections were [and are] indeed acquired as free donations.'[24] As we find in one historic parody of museum donation,

98

It is my wish, it is my glory,
To furnish your knick-knackatory,
I only beg that when you show 'em,
You'll tell the friend to whom you owe 'em.[25]

By accepting the gift, the curator flatters the donor and becomes beholden. The museum then pays for the upkeep and storage forever: gifts are not free, especially on the scale of some technological artefacts.

This is even more complicated when the material is not given but lent. Beyond the circulation of material between owners and museums that provide content for exhibitions, there lie in museums many items that still belong to someone else. Even the mighty Concorde, perhaps National Museums Scotland's most famous artefact (aside from the remains of Dolly the sheep), is actually borrowed from British Airways. Loans can be surprisingly long term: the stunning George III Collection of Scientific Instruments (illus. 48), commissioned in the early years of his reign from the instrument maker George Adams, is actually a loan. In the 1840s Queen Victoria had given this, her grandfather's pride and joy, to King's College London (which seems appropriate), which in turn lent it to the Science Museum in 1927. It has been the jewel in the crown of the instrument collections there ever since, subject of a sumptuous catalogue; some items tour the world as the 'Science and Splendour' exhibit, others are on display in the 'Science City' gallery, which we will explore in the next chapter.[26] And on an even larger scale is the loan of Henry Wellcome's medical collection from the Wellcome Trust (the body that manages his vast bequest). The trust has its own museum, the Wellcome Collection, but technically it still owns fully one-quarter of the Science Museum's holding. (Admittedly, this is a loan that is unlikely to be recalled at short notice, given the Wellcome Trust does not have anywhere to put it.)

Digital Collecting

Whether borrowed or owned, it is unlikely many science museums will take in massed material on Wellcome's scale again. But there are other routes by which they may grow their collections.

In 2011 the tech start-up company Bloom launched Planetary, an app that visualized a user's digital music collections in astronomical terms (artists as stars, albums as planets and so forth). It may not have changed the digital universe, but it did herald a quiet innovation two years later when the Cooper-Hewitt Museum, part of the Smithsonian Institution, acquired the code behind it. 'Preserving large, complex and interdependent systems

48 Exquisite instruments from the George ɪɪɪ Collection in the Science Museum's 'Science and Splendour' travelling exhibition at the National Science Museum of Korea.

... is uncharted territory,' remarked the curators at the time: 'trying to preserve large, complex and interdependent systems whose only manifestation is conceptual – interaction design say or service design – is harder still.'[27] Undaunted, they set out to treat the software like an animal in their sister organization, the National Zoo: they cared for it in the gentle environs of the museum's digital space rather than the online wilds. They also printed out a copy of the code, just in case (although this is no longer their practice).

Collecting software remains a challenge. Some museums do it well: among the 140,000-item catalogue of the Computer History Museum in the heart of Silicon Valley are over 16,000 instances of software (they are also world-leading in collecting associated paraphernalia such as punchcards and printers). For the Science Museum's antibiotics exhibition 'Superbugs: The Fight for Our Lives', medical curator Selina Hurley wanted material to represent how scientists measured the efficacy of antibiotics in pigs (the data can then be used to address antimicrobial resistance for humans).[28] Accordingly she collected a porcine cough monitor launched by the Dutch company Fancom BV in 2011 (illus. 49). Collecting the microphone and hardware was not enough, however; what had enabled the research was the algorithm that compiled the vast database and the software that analysed the data. And so she acquired these, and the accompanying manual, too.

Too often, however, the act of collecting digital entities is clumsy. Like many curators' offspring, my son reached museum saturation at a young age. But his young ears perked up when my National Museums Scotland colleague Alison Taubman mentioned that she had included the videogame *Grand Theft Auto* in the gallery 'Scotland: A Changing Nation' because it had been developed in Dundee. After some searching, we found this on display – at which point his eagerness turned to disappointment followed by incredulity. The game was not playable, as he had hoped, but rather was represented simply by the disc in its case. Despite being 'the greatest disappointment of [his] life', the game box does serve a useful function in drawing attention to the challenges posed by collecting digital things. Like a musical instrument in a display case, the museum collected this game but then *silenced* it. If the Smithsonian captured *Planetary* for a digital zoo, National Museums Scotland hunted down GTA and taxidermized it in a display case.

Whereas archives seek to preserve electronic data, science curators are interested in the tools and structures that generate and surround them, and they have been squabbling about

49 A pig cough monitor on display in 'Superbugs: The Fight for Our Lives' at the Science Museum.

different approaches since the 1980s. Science museums have never been very good at collecting time-bound media: they collect telephones, for example, but not the conversations they once channelled. Similarly, they collect the digital tools that afforded communication and other practices, even if not an archive of the communication itself. How can science and technology be collected in the 'age of algorithms'?[29]

103

The phenomena in question are *born-digital*: 'objects that take shape on a screen or hide in the back end of a computer program, composed of data and metadata regulated by structures or schemas'.[30] There are such entities relevant to all museums – art and design museums like the V&A lead in this area by acquiring digital art – but for science and technology museums, there is a double imperative. Digital things are important not only because museum visitors are interested in the development and use of technology over time in general, but because so much of contemporary science depends upon information technology. While video games and memes are clearly important to collect as a cultural record, technology collections' custodians need also to think about born-digital beyond personal use, to consider algorithms and software in other realms, from engineering to the military.[31] Software and databases are as much tools of the scientific enterprise as microscopes and test tubes.

This may seem daunting, but curators are not deterred. Collecting the born-digital has its own particular challenges, but so do other forms of acquisition. Yes, there is a profusion of material from which to select, but as we found earlier this is also the case for physical objects. Yes, it demands resources and advanced skills, but so do other sorts of collecting. Yes, historic entities such as early software require expensive maintenance of the original hardware or else complex simulations, but it is also costly and difficult to store and maintain steam engines and particle accelerators. Yes, there are thorny intellectual-property issues, but museums have always wrestled with these.

Rather than exceptionalize this burgeoning element of museum work, then, it should be embedded in curators' practice. Collecting software and hardware together, like the Science Museum's pig cough monitor, is the way ahead. Hardware without software is mute; software without hardware is trapped. Every collecting act is also a multimedia exercise involving information (texts and images) alongside each object. Even though I indulged here in the bad habit of isolating digital collecting in a separate section, curators need to get away from our habit of thinking about digital collecting as distinct from collecting physical objects. In collecting digital information and things alongside other media, science collections would do well to move towards the 'post-digital' approach advocated by digital museologist Ross Parry and others, whereby these media are integrated within other elements of museum work.[32] Digital objects and assets should be collected alongside material, visual and textual entities, just as the pig cough monitor was part of a wider effort to represent antibiotic efficacy.

Uncollecting

There is one more thing to unlock in this chapter, for as the director of the Guggenheim Museum Thomas Messer once observed, 'like any individual, no museum can consistently ingest without occasionally excreting.'[33] If museums collect and collect ad infinitum, even at the more modest levels outlined above, they will eventually fill any building they might have (even the vast hangars we will encounter in the next chapter). And so museums do sometimes get rid of things by (euphemistically) 'deaccessioning', (counter-grammatically) 'de-growth' or (more brutally) 'disposing'.[34] Botanical metaphors proliferate: some refer to 'pruning'; I prefer 'weeding'. Rarely does this involve actual destruction, but rather ridding themselves of redundant or defunct material by transferring items out of the collection to a good home, reversing the acquisition routes laid out above.

Deaccessioning is not new. Already in 1913, in the days when most things in collections were on display rather than in store, the Science Museum had a committee to assess material deemed unnecessary for exhibiting; of the 14,340 items acquired by the Science Museum in the 1920s, 4,637 have since been disposed of (or returned) according to strict self-imposed criteria.[35] Of all the items ever accessioned by the National Museums Scotland Science and Technology Department, nearly a third have since been deaccessioned for one reason or another.[36] During its decorously dubbed 'Collections Reform Programme', the National Maritime Museum in Greenwich tackled this on a massive scale, largely through a distributed network of other maritime museums.[37] The port engine of the steam paddle tug *Reliant*, for example, was surplus to requirements at Greenwich but the Markham Grange Steam Museum was happy to accept it to periodically run it for enthusiasts 'under steam' (actually electric power). Back in Scotland, a specialist group of industrial museums assessed their tool collections and sent duplicates to artisans in Tanzania and Sierra Leone via the charities Workaid and Tools for Self Reliance.[38]

Across the Atlantic, disposal is a more integrated element of collections management, and selling is more acceptable. The u.s. National Air and Space Museum keeps a list of possible transfers for swapping with other museums; 242 items were available when I last checked, including missile nose cones, 65 parachutes and a 3.5-metre Junkers Jumo turbojet engine.[39] The Henry Ford Museum of Innovation near Detroit sold 28,500 duplicates in 2000–2002, using the funds to buy more collections. In Ottawa, as part of the collections reorganization necessitated by the new collections centre, Ingenium deaccessioned swathes of its collection in a 'Collection Rationalization Project', first by offering to other Canadian museums, then only selling as a last resort.[40] Like a house move, a collection relocation is a good time to get rid of stuff.

Disposal is part of good collection management, then: but it is not the easy option. It is painstaking work – 'unreasonably onerous', opined one curator – and ironically, in the end curators often know more about deaccessioned items than others in the collection because of the care taken in the background research before transferring out of the collection.[41] This is the paradox of disposal. Still, museums accession far, far more than they deaccession, which is not a sustainable situation, as we will see in the following chapter. But done well, deaccessioning is key for science collections, and indeed museums are morally bound to do so. They can only survive and thrive if they undertake sustainable collecting: actively purging redundant collections, and acquiring fewer, more impactful objects with their entwined stories.

Collecting Stories

As they arrive and depart, science objects shape science collections. The other activities laid out in subsequent chapters notwithstanding, collecting is at the heart of what the museum does. It is therefore important that the mechanics of this process be clear, not only to you as readers, but to all those interested in museums, especially those collections that benefit from public funding. Science curators collect differently to their peers in other disciplines, to their predecessors and, no doubt, to their successors. They may no longer have grand encyclopaedic ambitions, but there is a hubris in selecting a small number of items from the profusion available. These are the things, words, images and digital entities that will represent the now for centuries to come.

This is a humbling thought, considering the role we found that serendipity and luck continue to play in the development of collections. It may, then, be reassuring to find that curators do not – or at least, should not – operate in a vacuum. Scientists have already made judgements about what is significant (as for example Canada's first laser, which was already considered important enough to be on display in the National Research Council laboratory before it came to a museum), and other forms of consultation shape collecting choices. Furthermore, museums have collection development strategies that harness these elements and focus energies into building on the strengths of a collection. 'Collection development' is an elegant catch-all term for collecting and disposing of objects, with a nice forward-looking spin. Strategies define a collection's regional, topical and chronological remit, but remain broad enough to allow for action when opportunity knocks, as when an important longitude clock comes up for auction, or a power station is on the brink of being demolished. Collecting is rarely an exercise in gap-filling, but rather a selective and tactical endeavour to strengthen a partial selection of the material culture of science and technology. To indulge in a geographic metaphor: science collections are not land masses with lakes that curators seek to fill in, but rather archipelagos, islands in the stream to build upon.

Given that there are many other ways of recording and experiencing science – on television, in books, online – why continue assembling material collections? It isn't just the tangible qualities of the objects acquired (or intangible in the case of digital collecting) but rather the lives these objects have lived, their provenance. A decent collection development strategy includes 'the astonishing stories of creativity and humanity embedded in the collection'.[42] Curators collect not to expand the already massive collection for their own sake, but rather to collect these stories.

Take the OncoMice we met at the opening of this chapter, two tiny mammals that encapsulate science collecting. They were material manifestations of the Science Museum's wider collecting around biotechnology. They are humble products of a massive global enterprise, powerful representations of research, of intellectual property, and of the ethical challenges of laboratory work. They are one link in an ongoing project, showing that science is an unfinished enterprise. They themselves seem everyday, two among a profusion of rodents in museum collections and laboratories, but the genes within them are complex. With these two tiny mundane objects museums can engage audiences with something as profound as the history of the ongoing search for the cure for cancer.

Likewise the TOF2 mass spectrometer, as well as embodying the history of physics in all its improvised, human glory, renders the conceptual tangible. As curator David Pantalony reflected, 'The beauty of collecting physics is that the most abstract of variables such as time and space become concrete, local and sensory.'[43] Physical black boxes help visitors to unpack the figurative black box of science: its tools and artefacts reveal its processes. These stories show that science is messy, contingent, human and social. The objects collected allow us to tell stories about concepts, about systems, but most of all, the objects tell us about people. Some are of scientific triumphs, some about the everyday life of scientists, doctors and the people who use these devices. Each and every object in the collection has gathered associations on its journey into the collection, and has continued to accrue meanings during its museum 'afterlife' as different people have handled it, written about it and experienced it – which is as much the case for items in store as for those on display, as we will see in the following chapter. Every generation of curators in their acquisitions lays down another stratigraphic layer in the collection. Every acquisition, from spectrometer to transgenic mouse, adds another story; and also seeds the potential for more stories penned by future curators and historians. 'When I am collecting an object', David feels, 'with all its 3D complexity of materials, markings, design, modifications, aesthetics, traces of making, symbolism, provenance, there is also a powerful latent potential for dozens of stories yet to be uncovered.' Which leads him to conclude that 'collecting is not completing a historical process, but starting one.'[44]

That curators collect stories has not been the only element of science museum practice revealed in this chapter. As well as re-emphasizing the human elements of science collections, we have explored the different routes by which objects arrive, each of them channelled by networks of scientists, doctors, dealers, programmers and others: kit brought back from a victorious field trip, an instrument given by a generous donor, a clock purchased at an auction, an algorithm arriving on a data stick. We found that science curators' 'field sites' are neither jungles nor excavations, but factories, power stations, hospitals and laboratories such as that from which the OncoMice arrived at the Science Museum. We have found that collecting is partial, rather than encyclopaedic. Finally, our detour into deaccessioning practices reminded us that the traffic of material culture is two-way: collections are dynamic entities, belying their reputation as dusty hauls of bygone eras. And as we shall now find, the action is not only in their exhibits, but behind the scenes in their vast stores.

50 The Mignon model 3 index typewriter.

3

TREASURES OF
THE STOREROOM

The typewriter seen here may look strange, different from any typewriter you know (illus. 50). Even if you are a digital native, however, you may have seen fetishized devices like this on the wall of a trendy coffee bar or in a smoke-filled room in a film. This is an example of an 'index' typewriter, one of the several forms that competed in vain with the QWERTY keyboard form on the laptop on which I type these words. The index typewriter user would pick out one letter at a time on the index plate by moving the indicator needle, pressing one of only two keys – the other would then insert a space. This was a surprisingly popular design at the time. Even though the era of index typewriters was waning, more than 350,000 Mignons alone were manufactured. This one, a model 3, was made in Berlin by Allgemeine Electrizitäts Gesellschaft (AEG) in 1914 or 1915. This machine was supplied in the UK by the Electrical Company before it was disbanded under the Trading with the Enemy Amendment Act 1916, and sold to Frank Smithies, a leading figure in the Edinburgh labour movement.[1] It later found its way to Brownlee & Son Watchmakers, who then donated it to the Royal Scottish Museum in 1970. It is now in the National Museums Scotland technology collection, along with sixty other typewriters and dozens of office devices from the decades around 1900, when typewriter technology was young.

The Mignon 3 is not usually on display, however, but rather is housed in an off-site collections facility at Granton, near the Firth of Forth, 5 kilometres (3 mi.) north of Edinburgh, one of millions of items in quiet spaces such as these, where staff and researchers care for them and find out more about them. To meet both people and things, here we will continue the previous chapter's journey behind the scenes, revealing the secrets hidden within this and stores like it. As well as the facility at Granton, we will explore other museums' depots in order to understand what is kept there, how it is looked after and how it is used. On our visit we will encounter other researchers, conservators and, especially, collection managers. Most of all, we will meet many, many objects, from a giant steam excavator to a delicate torsion balance. They are part of the secret bulk of the iceberg that sits below the public surface, and as we explore, we will find out what these people do with these objects in these remarkably lively places.

Storing Collections

Science collections facilities are extraordinary spaces. In an unassuming building, the main area for technology at National Museums Scotland – which for security reasons we shall call 'Building F' – houses huge scientific instruments and industrial engines. Such massive, charismatic machines are often what first catch visitors' eyes. I too was astonished on my first visit to Building F, despite twenty years of visiting museum storerooms. I too gazed in awe at the giant machines looming over me; I was especially taken by the tallest object of all, a steam excavator (illus. 51). It is one of the larger objects in the collection; we would need to take the wall off the building to get it out. One of the last of its kind, it looms strangely – now an unfamiliar object, it is curious and wonderful. 'Wonderful', 'awesome' and even 'magical' are indeed common responses to science and technology stores.[2] And yet even though more information is now available about stored collections online and in print, it remains a surprise to most people how many museum objects are off display.

The need for these facilities makes sense when we consider the sheer quantities of objects involved. There are no recent data for the overall picture in the UK, but it seems likely that around 7 million of over 200 million physical objects in museums may be categorized as science, industry, transport or medicine (natural history and archaeology make up the bulk of the national tally – but most of these things are very small). In U.S. museums there are around the same number of these kinds of objects, although the national total is much larger, at somewhere between half a billion and a billion items.[3] Individual science and technology collections tend to number in the tens of thousands: the Museo della Scienza e della Tecnologia in Milan has 18,000 artefacts; Glasgow Museums have 21,000 transport and technology objects; the Franklin Institute in Philadelphia has 43,000 items; and the Science & Technology holdings at National Museums Scotland comprise up to 80,000 things, depending upon what is counted. There are then a small number of very large collections. First among these behemoths, the Science Museum in London displays only 12,000 of its 425,000 artefacts; when it was last fully open, the Deutsches Museum displayed a whopping 28,000, over a quarter of its holdings; Moscow's Polytechnic Museum's total collections number over 229,000 things. These figures will never be precise, given how tricky it is to work out what it is being counted. An instrument such as the compound microscope made by John Benjamin Dancer at the History of Science Museum in Oxford comprises 350 items if its constituent parts and accessories are counted (illus. 52). And as we will find during our exploration here, two-dimensional items outnumber artefacts many times over (when including paper, photographs and other flat things the Science Museum Group collection numbers in excess of 7.3 million).

Such large collections need large spaces. A sizeable facility such as the London Transport Museum Depot has 6,000 square metres (64,500 sq. ft) of floor space, around the size of a soccer football pitch. Science & Technology stores at National Museums Scotland are spread across the Collections Centre and the National Museum of Flight and occupy over 7,000 square metres (75,000 sq. ft). But this does not account for shelving or stacking, or racking, or the height of the rooms. (The science curator measures in

sheer volume: Transport and Technology collections in Glasgow occupy 25,500 cubic metres and their equivalent at National Museums Scotland 33,000 cubic metres – more than five Goodyear blimps.) Far larger are the storage areas spread across the sites of the Science Museum Group and their equivalent collections of the Smithsonian, as we might expect – both in the order of 50,000 square metres (538,000 sq. ft). The flagship of the former is the brand-new Building One in the National Collections Centre on a former airfield in Wiltshire, 130 kilometres (80 mi.) west of the Science Museum in London. Building One has a footprint of 20,000 square metres (215,000 sq. ft) with a further mezzanine level of 7,000 square metres (illus. 53), with more than 1,000 square metres

51 Sleeping giant: a steam excavator built in 1926 by Ruston & Hornsby Ltd, an engineering firm in Lincoln, which built roads until 1958 (including the A8 from Edinburgh to Glasgow). The excavator languished in a depot until 1984, when it was acquired by the museum. Disassembled and carried on large-load trailers, it was painstakingly resurrected in the largest of the storerooms, its tip reaching the rafters of its home at the National Museums Collection Centre. It is 6 metres (20 ft) high and 8 metres (26 ft) wide, weighing 25 tonnes.

(11,000 sq. ft) of crucial support spaces – areas for study, conservation, photography and hazardous materials, all with enviable 'functional adjacencies' – to care for and study the 300,000 items housed there.[4] Another shiny new facility is the giant 36,000-square-metre (387,500 sq. ft) 'Ingenium Centre' built in Ottawa to house 90 per cent of the 85,000 objects from the Canada

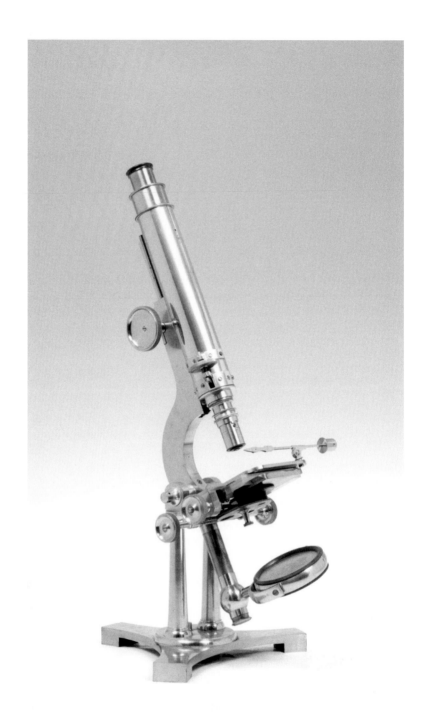

52 A compound microscope, *c.* 1860, by John Benjamin Dancer, Manchester:
this elegant instrument comprises hundreds of pieces.

Science and Technology Museum next door, as well as its sister aviation and agriculture museums, including twelve locomotives, 177 road cars and 190 bicycles (illus. 54). Like the UK's National Collections Centre, it will also house 2 million two-dimensional items and boast study facilities, including a research institute and a 'digital innovation lab'.[5]

Few organizations are lucky enough to have large new facilities like these; most others find space where they can, in basements and out-of-the way spaces, down pipe-lined corridors or in repurposed warehouses off-site. But whether improvised or bespoke, few are formally termed 'stores' any longer. The Smithsonian boasts the vast 'Museum Support Centre', and generally uses the neutral term 'collections space'. The Museé des arts et métiers' bespoke collections facility in suburban Saint-Denis is simply 'Les réserves'. Glasgow has a Museums Resource Centre with 'pods' and a new facility at Kelvin Hall is anything but a museum store. This giant former exposition building has been converted to house the research collections from the Hunterian (the University of Glasgow's museum), films from the National Library of Scotland and a fully equipped gym. These are active, multifunctional places.

Accepting that 'store' remains a shorthand, back in Scotland let us return to the Building F store. It is not straightforward to access, given that the guardians of these treasure troves are – rightly – highly security-conscious. Building F was completed in 1996 on the site of a post-war training facility turned over to museum storage in the 1970s; also within the perimeter are some of the original

53 Architect's impression of the Science Museum Group's vast Building One at the National Collections Centre at Wroughton, Wiltshire.

54 The Ingenium Centre, which holds most of the collections of Canada's Museums of Science and Innovation.

prefabs. To enter this National Museums Collections Centre we first check in at the gatehouse. Entry criteria are necessarily far stricter than for the public areas of museums. Access is rigorously policed, and most stores will involve a staffed passage point with dedicated security personnel – literal and metaphorical gate-keepers – carefully logging all visitors, entering at prearranged times with proof of identification. When I visited the vast collections facility of the Polytechnic Museum in the Tekstilshchiki District of Moscow, I pre-booked and went through the rigorous vetting procedure, showing my passport and posing for a photograph.

In Granton, as in Tekstilshchiki, a visitor would be met by one of the locals: a collection manager. This may be their formal job title, or else this may be part of a curator's role. Either way, they take their responsibility for the collection very seriously, and they are a house-proud people. Entry into a museum collections facility will usually be accompanied by apologies for some perceived untidy sin or other. (The one exception in my experience was

Katherine Ott of the Smithsonian, who was rightly proud as she unlocked the impeccable medical sciences storeroom in the National Museum of American History.) Once inside, apologies are redundant, difficulties and complications of entry are soon forgotten. Past the steam excavator in the large object room of Building F are stationary steam engines, an electricity generator, a vast bubble chamber for tracing subatomic particles and the flare tip from the Murchison oil platform. Perhaps most striking is the 6-metre (20 ft) Tod Head lighthouse lantern, painstakingly reconstructed, not only for its beauty but because putting it back together again reduced the footprint of the many crates in which it had been housed.[6]

Walking further into the building, we enter a space redolent of the final scene of *Raiders of the Lost Ark* (1981) in which the eponymous chest is wheeled down an anonymous aisle flanked by crate

55 The giant collections facility of the Polytechnic Museum on the upper floor of a huge former factory, showing good crating practice.

56 Shelved objects in the Harvard University Collection of Historical Scientific Instruments. Photographs of them adorn the vinyl covering on the window of the public Putnam Gallery, at once providing uv light protection and a taste of the storerooms.

upon crate – a visual metaphor for concealment. Crating is one approach to containing and protecting larger objects (illus. 55). Glimpses within these giant boxes can be rewarding. In Building F we notice a large crate has had its lid temporarily removed to provide access for a researcher, so peeking over we are greeted by the arresting close-up sight of a 3-metre version of HMS *Terrible*, a steam-powered warship, which even at 1:64 scale is impressive. Model ships are mainstays of transport museums and given their size and fragility they are often stored in crates.

Most objects are smaller, however, and their storage is arranged accordingly. Without the need for such Herculean floor-loading, these can be upstairs, and so upstairs we will go. Ceilings are lower, and collections packed into serried racks of shelving (illus. 56). We can get very close to the objects in storerooms like this: to cameras and slide rules, model steam pumps and microscopes.

Silent audio-visual collections include gleaming 1950s televisions, videodiscs and other long-forgotten formats. One bay is devoted entirely to oscilloscopes; another glass-fronted cupboard reveals prototype prosthetic limbs. Flatter and smaller objects are kept in drawers lined with polyethylene foam (the collection manager's best friend); as are the immaculately arranged valves in the Harvard Collection of Historical Scientific Instruments (illus. 57). Photographs, manuscripts and maps are lovingly arranged in clear polyester sleeves.

Building F is redolent of many other large bespoke stores around the world. Spatial similarities include the way they lend themselves to photographs with a distant vanishing point at the end of a rack (illus. 58). For medium-sized objects at least (illus. 59), adaptable steel-strip racking and the grey aesthetic of twentieth-century laboratory equipment are commonplace. Some racks are on rollers, compacting the storage by removing the need for aisles (although this does then increase the floor loading). Elsewhere,

57 A drawer of valves from the Harvard University Collection of Historic Scientific Instruments, carefully arranged in bespoke packing foam.

121

58 Brand new collection storage at Kelvin Hall in Glasgow, showing black plastic 'bread trays', a handy way of neatly accommodating small objects. This vanishing-point composition is a common feature of photographs of museum storerooms.

storage units are recycled or adapted, as at the Franklin Institute, where the collections took over former library shelving.

One feature of many of the objects on these shelves are the small paper tags attached or near them (some even inscribed on the objects themselves). The Mignon 3 is labelled *T.1970.45*: 'T' for technology, followed by the year it was accessioned – a common practice across collections. Its museum-ness is manifested by this unique number, which provides a thin conceptual line between order and chaos. The numbers correspond to their entry in the museum's catalogue and often a file stuffed with associated information. Originally scribed in ledgers, then printed on catalogue cards, and now entered in elaborate digital collection management systems, numbering and cataloguing museum objects is a time-consuming, technical and never-ending process. An object without a number is forlorn. Some are awaiting them: they comprise the bane of collection managers, the cataloguing backlog. These undocumented holdings are a dirty secret of the

59 Medium-sized 20th-century laboratory equipment in a science collection storeroom.

museum sector: even though they are dwarfed by the backlogs of natural history museums, tens of thousands of items in even some of the more renowned collections that feature in this book are yet to be formally catalogued.

In whatever medium, an important function of the catalogue is to connect the object to relevant records and images. We see many of these as we walk through the stores, in filing cabinets and shelves, in leftover spaces between racks. Here is a run of auction brochures with valuable contextual information; there is a pile of correspondence related to how an object was acquired. Further along we find a collection of stunning lantern slides and some vividly coloured mining maps. For technology collections in particular, engineering drawings and instruction manuals can be very important, showing how these still, quiet things functioned during their working lives (illus. 60). The Science History Institute, for example, houses over 5,000 manuals accompanying its rich instrument collections. In Building F, as well as the baggage label that identified Smithies as its former owner, the Mignon 3 is accompanied by a valued but often accidental accumulation of instructions and the like. This supporting paper and print, alongside physical collections in store, forms what we might call a 'penumbra of documentation', invaluable but overlooked.

Paper manifests in other important ways in the work of collections facilities. Print-outs, posters and signs adorn the racks and walls of storerooms, some bearing images showing what is within boxes, others listing particular collections, yet others alerting collection managers to a stiff roller rack. And among them in Building F we see some especially vivid warning labels: for we have come across the radioactive-material store. This segregated room holds material too dangerous to keep out in the open, but too precious to remove – science collections staff have the dubious honour of being the museum professionals most likely to keep a Geiger counter in their pocket. Storerooms may also have dedicated areas for chemicals, explosives, firearms, human remains or narcotics. Plastics discolour, warp and fray; they can even infect others as they 'off-gas'. Asbestos, used widely in heat resistance

60 The 1976 computer board for original Apple 1 personal computer at the National Museum of Scotland, which makes little sense without its manual.

and strengthening, but now known to cause pulmonary diseases, is a particular bane of the science and technology collection. (The scale of this in science museums dawned on me when I arrived at National Museums Scotland and asked a colleague what proportion of the department's collection contained asbestos; she indicated the question would be better framed as how many did not.) For the most ill-behaved objects, licences are held, audits taken. But the risks inherent in some of these objects never cease to surprise: some antiseptic gauzes found in medical collections contain picric acid, which when dried out can, alarmingly, explode upon movement. Thankfully, conservators generally identify the risk before the bomb disposal squad needs to be brought in. Risk management is a core element of what museum professionals do in stores.

Working (on) Collections

Fortunately, the Mignon 3 has no asbestos; had it been a recent arrival, this would have been confirmed by another inhabitant of Building F, the conservator, in the next space we will visit. While other disciplines might have a conservation studio or laboratory, the science and technology conservator has a workshop. There we find lovingly maintained machine tools; electric gadgets for keeping the push-button attractions in the museum galleries running; all manner of screwdrivers, hammers, brushes and safety equipment everywhere. There is a common assumption among other areas of the profession that just because science and technology collections appear to be robust, they need no environmental controls and little active care. This attitude, and their sheer size, mean that some are left in unsuitable storage conditions – or even worse, outdoors – where they rust and degrade. On the contrary, they do need attention, delivered by a dedicated cadre who clean, stabilize, maintain, preserve and restore these peculiar objects.

A lot of preventive work goes on beyond the workshop: controlling the environment in the store, managing temperature, light and relative humidity. But some objects demand further interventional and proactive work, a range of activities that come under this umbrella term 'conservation'. It can have different aims according to whether the museum wants the object to function and/or recapture its original form. Without its movement, is a machine mute and dead? Should we make it work again? Should we seek to restore it to how it looked when new, when used or when it was transferred to the museum?[7] Answering these questions involves selecting the appropriate treatment path for a given object. To preserve is to treat it just enough to arrest deterioration, to maintain it sufficiently for it to survive for the posterity we ask of it. The mantra of the conservator is that, as far as possible, processes applied are reversible, and exhaustively documented (illus. 61). Restoration, by contrast, involves bringing the object back to its original condition. This will involve removing later accretions, even scraping away the efforts of previous conservators. If a significant proportion of new material is needed, spare parts and the like, this may tip treatment into reconstruction, often to bring an object to its full working glory. This could involve rebuilding certain parts or adapting them to ensure they are safe to use. The logical extreme, finally, is complete replication.

In the Building F workshop, for example, we meet Sarah Gerrish, a conservator who specializes in wooden artefacts and furniture – encompassing carts, aircraft wings and, especially, clocks. The latter are a very particular and demanding breed of objects that can be found in art, design, social history and technology collections. Whereas much of the work on clocks focuses on their intricate inner workings – the work of other highly specialized horological conservators – Sarah treats their visible shells. As a conservator she is a professional problem-solver, piecing together unidentified fragments and improvising solutions. Her work spans the fine line between remedial conservation and proactive restoration. Her day-to-day tools include a scalpel, tweezers, hammers, many clamps and a mallet she has modified by taping cushioning on one end, so she can tap with varying pressure (illus. 62). Her work depends especially on adhesives: epoxy resin

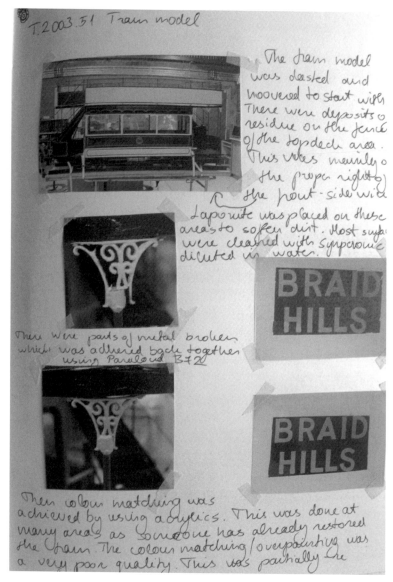

61 Conservator Julia Tauber's notebook, painstakingly recording the treatment of a tram model.

filler, wood glue and fish glue – first aid for objects. She uses HMG Paraloid B72 glue, for example, which sets slowly to allow her to adjust the pieces she is working with, then cleans off the excess with acetone: endless paper towels are also part of her toolkit. Yards of green non-tacky tape keep pieces in place as they dry.

Like the researchers we will meet below, she also needs light, so she takes her own spotlight with her.

Sarah worked at national museums in Wales and Scotland before going freelance, for no museum has all the skills it needs in-house. Much museum conservation is undertaken by independent contractors, or otherwise by enthusiast groups that have a wealth of skills and experience, often from the industry that originated the objects, such as aircraft engineers or computer programmers. Sarah's collaborative approach also illustrates that conservation is almost always teamwork of some variety, given the diversity of size, function and material in any collection: wood conservator and metal conservator, clocks workings or shell, internal and external, amateur and professional. There is no single 'science conservator'. In the preparation of material to go on display at the Riverside Museum in Glasgow, for example, 38,000 hours of conservation time were spread across those with expertise in natural history, fine art and transport. Conservators involved remembered working 'on objects from across the collection, from steam locomotives to toy cars, paintings to ship models'. The rigging on one of the latter was 'literally being held together by dust!' Addressing these varied challenges was a multi-skilled team effort, 'from a variety of areas such as fixed and temporary

62 Conservator Sarah Gerrish's tools for working on the wooden parts of clocks.

63 Challenging collection moves: Aerospace Bristol moves a Sea Harrier with the help of an RAF Chinook helicopter, 8 March 2017.

contracts, conservation packages, volunteers and apprentices from outside industry'.[8]

What unites curator and conservator, technician and scientist, amateur and professional, is a shared passion for the artefacts, a bond with their machines, what anthropologist Sharon Macdonald terms 'object-love'.[9] In watching work in museum storerooms, Macdonald also astutely observed that 'their guiding motif is a quest for order'.[10] This is most visible in the amount of object movement that goes on. Contrary to their reputation as static mausolea, work in storerooms in general involves frequent object transfer. As Sarah finishes up with her clocks, and wheels a trolley of pristine hoods back to the horological storeroom, we notice she is not the only trolley pusher in Building F. Few conservation or cataloguing jobs can be done to an object *in situ* on its shelf; they are moved around from rack to rack, to conservation work-shop or to external contractor. Most objects are moved by trolley; certainly this is how the Mignon 3 and other typewriters travel. The rhythm and day-to-day existence of museum storerooms are governed by trolleys, their availability and the routes they can follow – or not: like Daleks, collection managers are vexed by stairs, which appear in awkward places in these facilities. At Kelvin Hall, for example, by a quirk of converted architecture, a staircase greets you as you enter the Hunterian's storage area. Larger objects, meanwhile, present still larger challenges. Getting them to stores may require hauling, towing, rail transport or specialist lifting (illus. 63). Trolleys may be the lifeblood within the museum premises, but some require forklift trucks, cherry pickers, winches and cranes.

Movement can bring order from disorder. Unlike pristine exhibition areas in which collections are impeccably arranged, storerooms usually have some objects out of place, in a temporary location before moving on to somewhere else. These things-in-limbo might have been acquired and not yet catalogued or shelved; their room or shelf might be being painted or otherwise worked on; the objects might be awaiting a researcher or a display. We might not like to admit it, but objects may have been out of place in this way for some time, even years.

Beyond this day-to-day movement (or lack thereof), period-ically a larger task faces curators and collection managers, when the contents of an entire room or building need to be relocated. This may be because the space is no longer suitable, or is needed for another purpose. This is not always well planned, with press-ing deadlines and few staff to meet them. Or worse: one horrified curator recounted the fate of a storeroom containing collections in their custody that was not emptied of its objects when build-ing work was undertaken; holes were knocked through the walls to allow ventilation of the site, leaving a layer of construction dust over the collection. It took two years and a lot of money to restore the collection to its former glory.

The Mignon 3, like many of the collections in Building F, has moved several times in the last century: from the main museum in Chambers Street to improvised facilities such as a former cus-toms house in the Port of Leith, before settling in Granton. Let us take a detour to other collections that have moved in recent years. In 2016 the History of Science Museum in Oxford moved its entire stored collection because another (richer) part of the university needed to use the Old Power Station in Osney, just to the west of the city, where the scientific instruments were housed. Staff took thirty months auditing, documenting, measuring, weigh-ing, condition-checking (including surveying for omnipresent hazards), photographing, cleaning and packing their 9,000 objects (illus. 64); in the process they found these comprised some 60,000 separate things. Up to eight staff were employed full-time, with help from volunteers, using barcodes – a common if intensive method of moving heritage holdings – and a seemingly endless supply of acid-free boxes and tissue, corrugated plastic and inert foam.[11]

The Old Power Station move was expensive and difficult, and compounded by a delay when the facility earmarked for the uni-versity collections was not yet ready, necessitating rented storage in the meantime. But spare a thought for the Science Museum Group in their move to Building One at Wroughton: the largest peacetime museum collection movement in the UK. Since the

64 The Move Project Team counting, photographing, cataloguing and cleaning the collection of the History of Science Museum in Oxford.

1980s, with the Victoria and Albert and the British Museum, the Science Museum had shared the former Post Office Savings Bank office building near Kensington Olympia known as Blythe House (famous as the location for the filming of the 2011 feature film version of *Tinker Tailor Soldier Spy*). The Science Museum's 12,000-square-metre (129,000 sq. ft) section included the bulk of Sir Henry Wellcome's 114,000-strong medical collection.

Back in 1980, when Blythe House was being prepared for occupation, the Science Museum also secured space for its larger objects in six hangars at a former military airfield at Wroughton in Wiltshire. It was some 130 kilometres (80 mi.) away with poor transport links, but Wroughton offered plenty of space. And yet, even though they were large enough that clouds can form inside them, by the 1990s the hangars were themselves full: airliners, trams, buses, tractors, rockets, the world's first hovercraft and a three-storey printing press take a lot of room; not to mention the 1.5 million books and archive materials that were transferred there in 2007. Despite repurposing and building other buildings on site, storage remained a headache over the following decades. Matters were taken out of the Science Museum's hands in 2014 when the UK government Department of Culture, Media and Sport prepared to put Blythe House up for sale, and all three tenant museums would be evicted. The blow was cushioned with a £150 million golden parachute between the tenants, and the Science Museum Group elected to restyle Wroughton the 'National Collections Centre' and transfer the Blythe House material as well as collections from their other sites.[12] The group initiated a programme, 'One Collection', to review their entire holdings, and move 320,000 objects to the vast Building One at Wroughton. Like Oxford's approach writ large, they used barcodes to track movement and the majority of the objects have been documented, photographed and published online as part of the process, diminishing their daunting backlog in the process.

Using Collections

Back in Building F, the Mignon 3 typewriter is rather easier to move: curators can carefully take it from its shelf and wheel it to a table should a researcher want to examine it. But why would they – what benefit does a direct encounter with an object like this bring? What does it mean to 'research' scientific and technical objects? Even if it is just one kind of research in one kind of space, this encounter with stored material is key. And yet here the category confusion we encountered in the introduction comes back to haunt us: the research that goes on in science museums is rarely scientific. Natural history museums provide important data and expertise for the study of evolution and biodiversity, for example, and archaeology holdings can be better used to understand human anatomy. They will sometimes use specimens for analysis, slicing off tiny parts to analyse DNA. But much of the research that goes on in the science and technology museum storeroom is more akin to that deployed in social history, art or anthropology collections. In these fields it is all too easy for curators or others to nod wisely when there is talk of 'researching' objects, as if they had a language that could be read like a text. As one science curator wrote, 'Looking hard at an object under review to make it "speak" is one of the museum curator's skills'; she refers to 'profitable "conversations" with such material'.[13] I asked one researcher whether the scientific instrument he encountered during group work at a seminar spoke to him: 'No,' he replied, 'but it spoke to the physics teacher next to me.'[14]

In communicating with an object like this, what does a museum researcher in science and technology museums do? How do they use these collections, which include some incredibly complex devices? Museum research, it transpires, can be surprisingly simple. To illustrate, let us walk through the steps involved in examining the Mignon 3 as a researcher. First, we need to find it. Thankfully, assistant curator Katarina Grant would be on hand to help. Even with that holy grail of curators and collection managers, the entirely accurate and up-to-date collections database, this is no mean feat among the tens of thousands of things on kilometres of shelves. The element of exploring and locating is a key, yet overlooked, element of what anthropologist Nicholas Thomas calls the 'museum as method'.[15] He draws our attention to the element of discovery, of 'happening upon', in a museum or its store. The things kept next to the item the researcher hunted for can often be revealing; and simple serendipity while browsing shelves can also be a powerful thing.

Shelves are difficult to reach and can be cramped, so to keep the objects safe and for our convenience the next step for a

65 Typewriter researcher James Inglis's haptic encounter with material culture: typing. This reveals things that the senses of sight, smell and hearing cannot.

museum researcher is to take the objects to somewhere more convenient and easier to lay them out and study them. In Building F we are lucky enough to have a dedicated room for this: what archivists call a 'search room', others a 'research suite', but at its simplest this is a space with a large table, good light and Wi-Fi. Most researchers are as likely to have a laptop with them as a notepad – if only to access the museum's catalogue, as we will no doubt have further questions of the object and others like it.

Once it is on the layout space, we measure the typewriter to double-check the data in the catalogue, and take some pictures. Our next and most important step is simply to look at the object for a sustained period, in close detail. An art historian might consider this connoisseurship – visual knowledge gained by looking at artworks, bringing visual memory to bear and by frequent comparison developing the capacity to recognize style, period, locale and authenticity, a crucial part of traditional curatorial expertise. We will look at decoration, iconography (any symbols or meaningful images on the object), fine details, the patina, evidence of how it was made and used (and even broken). The next stage for many researchers, if it is safe to do so, is to handle the object. For this haptic experience of a smallish metal object like this, researchers will carefully don purple latex gloves. Other researchers will be sporting characteristic white cottons, the material of the object dictating the nature of the handwear. Debates about the value and variety of gloves rage within the sector: for my insistence on the use of gloves I was once accused of being 'pathetic' by a revered television archaeologist and knight of the realm.

The first act of handling is to carefully lift the object. This helps us understand the balance of the machine, how cumbersome it is, and crucially for typewriters, its portability. 'There is an experiential aspect to portability that cannot simply be described with descriptions of an object's dimensions,' one typewriter expert reflected. 'Dust and dirt can get on your hands and clothes from carrying the object, which tells you more still about the object's condition [and] materials of construction.' Weight and balance are a particular issue with typewriters: 'some machines can be very top-heavy, the heavy carriages sliding around back and forth when moved which makes transport even more difficult. Are there points on the typewriter that can be easily handled or gripped?'[16] One heavy Oliver typewriter, for example, sports handles to aid lifting – although as any conservator will tell you, one never lifts a museum object by its handles. Even such robust objects can become fragile as they age, and this instils a powerful sense of caution. But it is worth it – handling the object can tell us what the closest looking cannot.

Some researchers will go even further and use the object for its original purpose. By using a machine, we can learn more about it than ever could be learned from reading or looking (illus. 65); this is especially the case for typewriters, which rely so much on speed and different finger pressure. Steven Lubar, while a

Mignon Typewriter, Model No. 3.

A. Space key

B. Printing key

C. Protecting cover or hood

D. Right Hand Spool

D¹. Left Hand Spool

E. Right Hand Screw Button for Spool

E¹. Left Hand Screw Button for Spool

F. Bell Trip

G. Right Hand Stop

G¹. Left Hand Stop

H. Platen Roller Knob

I. Paper Release

K. Line spacing lever

L. Switch for line spacer

M. Platen roller release lever

N. Right carriage release lever

N¹. Left carriage release lever

O. Right Lever for Rod (P) which grips the Paper

O¹. Left Lever for Rod (P) which grips the Paper

P. Rod for gripping the Paper

Q. Ribbon

R. Type Cylinder

S. Screw Nut for Type Cylinder

T. Metal Guide Plate

U. Scale

V. Carriage Rail

W. Keyboard

X. Pointer

Y. Handle for the Pointer

curator at the Smithsonian, learned more from a short spell with machinists trying to get a pin-making machine from the 1830s to run than he ever could have from written sources.[17] Operating objects, whether a pin-making machine or a typewriter, can also reveal the sound – and even smell – of these otherwise still, silent things.

One method of learning how to use the object is to consult the manual. Thankfully, given that we are now more used to QWERTY keyboards than index typewriters, the museum has an instruction booklet for the Mignon 3 (illus. 66). This re-emphasizes the value of the manuals we passed earlier in our tour, and of the penumbra of documentation more generally. Close study and comparison

66 The penumbra of documentation: London Electrical Company, *Instruction Book for the Mignon-Typewriter No. 3* (c. 1914), facing p. 12.

with printed sources are common elements of history of science and technology. Other important sources include the catalogue entries relating to the object, contemporary articles relating to objects like this, and photographs of other similar typewriters in other collections. This documentation and imagery help him to understand the objects' value, and its provenance – that is, the history of its ownership, its origins and route to the collection. This value is financial, of course, but also cultural and technical. Its provenance is important to understand how we came to have the object and the people involved, whether inventors, users or collectors.[18] That it travelled from Germany to Britain during the First World War and was then used by Smithies is fascinating and brings new meaning to the Mignon 3; other objects' provenance reveal colonial roots or dark dealings with dodgy dealers. The paperwork also helps researchers to piece together where these objects were used and by whom, and where they have been in the century since, for their museum afterlives are commonly longer than their use-lives. Words and pictures can help connect the biographies of things to the biographies of people, and thereby help us to understand not only the role of material culture in science and technology but the construction of their heritage in museums.

The researcher's next step is to select similar objects from storage and lay them out next to each other. Just as archaeologists will assess an entire assemblage, so in technical collections dozens and even hundreds of similar devices can be studied at once. Think, for example, of the manifold instruments collected by Henry Wellcome at the Science Museum. The comparative analysis afforded by such massed ranks of objects is characteristic of museum research; Nicholas Thomas terms this 'juxtaposition' in his museum-as-method.[19] We can see small variations, judge the chronological development of a machine and note differences between makers. This is the value of these vast assemblages, helping researchers to be able to make sense of individual things within the array. Museum research is relational: the objects are interesting not only in and of themselves but in their connections to other things in the collection and beyond.

Typewriters are just one kind of object, of course. In the looking/using/comparing approach we have taken here we are following a path laid down by others with different objects. They ask, where did it come from? What is it made of? How was it designed and made? How was it used? Together these queries have been grouped under the term 'material-culture studies' by researchers. By way of contrast with the Mignon 3, consider the exercise by historian of science Katharine Anderson of York University in Toronto and colleagues, who spent a week in the

67 Curators and other historians of science seek to understand an Eötvös torsion balance in the storeroom of the Canada Science and Technology Museum as part of the Reading Artifacts Summer Institute. They learned things from the object that printed and visual sources would not have revealed.

stores of the Canada Science and Technology Museum (in the days before the Ingenium Centre). With the guidance of curator David Pantalony, whom we met in the previous chapter, they set out to see what material-culture methods could tell them about a pea-green metal instrument of almost humanoid form (illus. 67).

They carefully measured and described it in detail, including its overall shape, a missing dial, and chips in the paint. They gloved up and carefully manipulated its moving parts. This confirmed to them this was a sophisticated geological precision instrument of some kind, and so they set out to find out more from textual and visual sources. They began with the 'trace' information they found on and around the instrument: labels and inscriptions on the packing case, engravings on the instrument itself. These revealed elements of its function (a torsion balance) and its Hungarian provenance. It was designed by Loránd Eötvös and made in 1928 by Nándor Süss, they read; registers told them it had arrived in the museum in 1987. Photographs in the lid of the instrument case set the design in its original aesthetic context, showing similarities between the instrument's design and that of the Josef Bridge in Budapest. Overlaid marks and tags also allowed the team to piece together the afterlife of the object.

Armed with these details the researchers set out to establish the context of the instrument in the published literature. This revealed its role in 1920s physics, in the attempts to correlate inertial and gravitational mass, which would in turn go on to support quantum theory. When they began to consider function, things began to get interesting, as contemporary descriptions of the balance did not quite match the artefact in question. History of science generally focuses on delicate laboratory-based experiments, but the robust form of this example pointed the researchers to a less-known fieldwork history, in this instance its use by the Canadian geologist A. H. Miller. Only by close inspection did the group find out that a particular light source was needed to operate the telescope embedded in the instrument. 'The dialogue between clues found in the literature and our limited attempts to use the instrument', they wrote, 'led us to the rediscovery of what had been tacit knowledge for the instrument's users, a detail left out of all the descriptions of the instrument, despite its essential role in the act of observation.'[20] The presence of the instrument in all its complexity also clearly showed that considerable skill and unwritten knowledge would have been required to operate it. The project left the museum with more knowledge of this artefact, the team with enhanced research skills and their students with a case study in material-culture studies. Each study, if fed back into the collection, wraps new layers of understanding and meaning around the objects.

Anderson and her colleagues are unusual among historians of science and technology. Many are interested in material culture, especially in light of an attention to things in scholarship that had been dubbed the 'material turn', but few will actually come to museums stores to experience objects directly. Noticing this absence in the stores with us, two of my museum colleagues surveyed the main academic journals and confirmed that very few university-based historians of science and technology use museum objects in their publications.[21] They may engage indirectly with them via the other media we touched on above – archives and books, for example – but they rarely encounter them directly.

If not academic researchers, then, who does visit stores to use collections? Most museums will go out of their way to welcome by appointment several dozen practitioners of one form or another into their stores each year (in the year I sampled, the Whipple had seventeen research visitors, National Museums Scotland Science & Technology hosted thirty researchers and the Franklin had 87 on-site queries). Students from a range of disciplines come to dip their toe into research, whether undergraduate or postgraduate; artists visit for inspiration; collectors to compare and situate their own material; genealogists to follow up a family history lead; hobbyists or 'expert enthusiasts' to enhance their own understanding. The Science Museum Group's 'Energy in Store' project, for example, brought together collection managers and former engineers to talk about the energy collections held by the group, which benefited both the museum and this user community.[22]

Crucially, these projects involved curators. Traditionally, research has been a core element of the curator's role, although this is not always the case beyond national and university museums, and even those who are research-active often give the impression this is a dwindling part of the job. Nevertheless, the Science Museum Group director, Sir Ian Blatchford, announced upon his appointment that he wanted to return to 'real research on our own collection'.[23] As well as curators, conservators also study collections, and visitor and learning staff research the relationship between objects and audiences. Some institutions choose to coordinate this research within a specific unit, such as the Science Museum's Research and Public History Unit, or the Research Institute within the Deutsches Museum. A department like this makes clear the intention that a museum can be a research driver (like a university) rather than simply a research resource provider (like a library).

Channelled through a dedicated unit or otherwise, curators and their colleagues will also facilitate others' research, whether as paid consultancy, hosting their visits to the stores or in response to remote enquiries. Enquirers are usually asking about the existence of, or further details about, specific objects, but their questions can range from the sublime ('where do you keep the spark that started the fire of London?') to the ridiculous ('do you have a life-sized cut-out of Nicolas Cage?').[24] Enquiries also vary in quantity between different institutions, of course, from fifty to one hundred a year by a small team like those at the Whipple or the Franklin, to the five hundred responses per annum provided by the Science and Medicine Division of the National Museum of American History in the Smithsonian.

These responses, which comprise a dialogue with museum users, are an overlooked but important collections research output for museums. Thinking about research outputs only in terms of specialist publications and their readers is to miss the many other ways that museum research lands (although I appreciate the irony that I write this in what may be viewed as a niche publication). Collections research fills catalogue entries for future curators. Collections research feeds into the successful object treatments for conservators. Collections research populates student work, from undergraduate assignments to doctoral theses. Collections research emerges in popular books for the general reader. Collections research enables documentaries and podcasts for remote audiences. Collections research feeds into the acquisitions we discussed in the previous chapter for future visitors and the blogs, tweets, online databases and webpages we will explore in the next. The breadth of museum research and its audiences is encapsulated nicely by interdisciplinary curator Martha Fleming, who has worked on exhibitions at the Deutsches Museum, Medical Museion in Copenhagen, the Science Museum and the V&A:

Museum research takes place in conservation and in curatorial departments, in learning and teaching, in exhibition-making and label-writing, in trend-forecasting and digital documentation – and in all the places in between. These collaborative encounters at the museum's core forge unique forms of knowledge (though they are not always visible to the public or seen as research by those involved).[25]

Martha helpfully points us to a major output of collections research: the exhibition. Every exhibition, from a one-case display to an international blockbuster, relies on research, often (but not always) carried out by curators. Every well-turned phrase on a text panel takes careful study and deft synthesis. This process of exhibition-making is distinct research aimed at textual outputs. The third dimension – the presence of things and their spatial arrangement – makes the exhibition markedly different in process and product from other kinds of research. Detailed specialist investigation is often necessary, and exhibitions are necessarily the product of teamwork, as we will find in the next chapter. The Mignon 3 featured in an exhibition at the National Museum of Scotland, 'Typewriter Revolution', thanks to a dozen-strong team of researchers, curators and exhibition specialists. Collaborative research projects between curators and university-based colleagues not only give rise to books and articles, but great exhibitions like this.[26]

The Norsk Teknisk Museum in Oslo has developed an especially collaborative way of working, facilitated by the 'LAB'. They seek to generate physical and conceptual interdisciplinary experimentation around things, with rooms for conservation (working on objects), research (working with ideals) and exhibit making (working up models).[27] The LAB method involves artists, scientists and other externals working with museum professionals in a series of seminars. One such workshop centred around a massive carved granite 'Hitler stone' that had been originally intended for a massive Nazi triumphal arch; following the LAB method an interdisciplinary team worked from this to the 'Grossraum' exhibition about forced labour and Second World War road building in Norway. We will meet the innovative exhibition-making practices of the Teknisk Museum again in the next chapter.

All these projects brought a variety of people into collections facilities: but still in the dozens rather than the hundreds. 'The specific research audience is numerically small but disproportionately significant,' reckons the Science Museum Group, 'as its members address otherwise unseen and unused collections and develop new narratives from them. Research sustains and helps grow our collection, uncovering forgotten relevance and revealing new stories.'[28] Some museums are seeking to provide access to these spaces more generally by rendering their facilities 'visible storage'. This vogue started in anthropology museums, and usually involves adding access to a high-density hybrid of storage and display alongside the more usual galleries we will meet in the next chapter. The National Railway Museum's 'warehouse', for example, is really a densely populated extension of the public gallery. Hunterian science curator Nicky Reeves calls this 'a carefully arranged performance of storage'.[29]

More significant is what we might call 'visitable' storage, taking groups around genuine collections facilities. One facility already with a track record in this respect has been the Glasgow Museums Resource Centre at Nitshill, 10 kilometres (6 mi.) south of the city. Its seventeen pods are intended to be visited, and the entire site is set up for learning groups: up to 15,000 attend annually. More organizations have intentions in this respect than realize them – it takes a great deal of care and staff time to ensure the safety of both collections and visitors – but both Ingenium and the Science Museum Group have ambitions for tours and teaching in their new facilities. Building F was originally designed with this purpose in mind, and plans are afoot to increase the rather limited public footfall. In the meantime, conscious that we have only touched on the hidden riches of this and other museum storerooms, let us leave the Mignon 3 behind.

Fragments of Eternity

We have now had a glimpse at what curators, conservators and researchers do beyond and behind the exhibitions we will meet in the next chapter (which are, after all, how most of us experience museums). Every day they work away in these multi-sensory spaces with the quiet sounds of their trolleys and the distinctive oil-and-heritage smell of science collections. There are many other people involved in the vast majority of off-display items as well, of course: security guards, porters, cleaners, collection managers, technicians, cataloguers and photographers who rarely step into the limelight.

Neither do most of the things they are caring for feature in the public profile. For although the Mignon 3 is destined for an exhibition, it is important to note that most stored museum objects are not necessarily waiting their turn to go on display. Some are too precious to exhibit, some are unshowable, some dangerous; many are very similar to those on display and to each other; some are just too mundane. And yet museums invest a huge amount in their care and upkeep – by my estimate, more than half of their resources. Museums are morally compelled – and in the case of many national museums, legally bound – to retain collections forever, which is a long time. This has practical and financial consequences that we too often ignore: at National Museums Scotland we calculated our real costs of acquiring, assessing, documenting and storing an object. Accepting that including an infinite time period would scupper our maths, we came up with a figure per decade, which was eye-watering enough. I will be discreet about the precise sum we arrived at, but it was on a par with a similar calculation made by an American curator who found it to be at least \$6,600 for every object for entry into the collection in its first decade. As one colleague reflected philosophically during our otherwise actuarial exercise, we are dealing with 'fragments of eternity'.[30]

Such costs are indicative of the main topic of this chapter, the great mass of material housed in these surprisingly lively treasure

68 We should celebrate storage: medium object storage housing the Harvard University Collection of Historical Scientific Instruments.

troves, which include millions of images and texts as well as material culture, cared for by a select band of mixed professionals and consulted by a handful of hardy researchers. The practices are rarely considered by museologists, let alone museum visitors, who tend, quite understandably, to focus on the objects on display.[31] But they are critical in our understanding of science objects, and in their role as cultural artefacts as well as technical tools.

Was it worth keeping the Mignon 3 typewriter for all these decades? Should we reinterrogate our custody of thousands of objects – hundreds of thousands – not all of them catalogued, a significant proportion of which, unlike the Mignon, will never be used? Are they all worth it? Yes, if we accept that objects' value lies in their potential for the other functions below. Yes, if we accept that storage may well be their ultimate resting place, and that the museum storeroom as a space, as a facility, also has merit in and of itself over and above being a holding bay for other spaces and functions. Yes, if we consider the store to be the material memory of science and technology. The many activities that go on there, the many things, pictures and texts, help us to understand science. Museums need not therefore apologize for the Mignon 3 and the other millions of items off display, but rather promote them, debate them, reveal them more broadly. Let us celebrate science collection storage (illus. 68).

69 Weird and wonderful: ferrofluid.

4

ENGAGING OBJECTS

n the tube pictured opposite is a ferrofluid, a liquid that acts like a solid in response to strong magnetism (illus. 69); suspended in the liquid are magnetic particles in the order of nanometres, the same scale as the graphene we encountered in Manchester in Chapter Two.[1] In normal circumstances the particles move freely and the substance is liquid, but under a magnetic field the particles form solid structures. Moving a magnet nearby prompts the ferrofluid to act like a cross between a hedgehog and a small, mobile storm cloud. As well as these eye-catching qualities, these properties have a variety of uses, for example in the magnetic ink used in printing banknotes.

Unlike most of the objects we have encountered in this book, this tube is not in a permanent collection, but rather is part of a temporary exhibition, one of dozens that were distributed around science museums in North America as part of a huge project to engage young people and their families with nanoscience through active hands-on learning. Spearheaded by the Science Museum of Minnesota and the Museum of Science in Boston, and funded by the United States National Science Foundation, the Nanoscale Informal Science Education Network (NISE Net) initially ran from 2005 until 2012. NISE set out to engage audiences with new nano-technologies via events, digital resources, training and pop-up exhibitions (illus. 70), involving over six hundred museums and other organizations, and reaching more than 30 million people.[2]

Later reconvened as the *National* Informal STEM Education Network, extensive evaluation indicated some success in changing both professional practice and public opinion around nanotechnology. NISE was in the end in receipt of $40 million in public funding. By parity, on this side of the Atlantic, the events offered by Ecsite, the European network of science centres and museums, attract some 40 million people each year.[3]

Science engagement is big business. The sites and practices we explored in the last chapter involved hundreds of thousands of museum objects, but only a select few people. By contrast, we will now consider the experience of the millions of people who visit museums, and especially how they encounter objects. The most visible way of engaging is via an exhibition, so we will start our journey there, before exploring other forms of engagement, especially events organized by museums and how they use digital channels to offer object experiences.

In all these activities, science museums are underpinned by a paradox: despite an abundance of information, many of us still

70 Exhibiting small: the 'Nano' exhibition at Port Discovery Children's Museum, Baltimore.

make anti-expert choices that confound scientists.[4] European and North American populations have access to more science than ever before via a range of channels: television, the Internet, news media, social media, zoos and science centres. We know that people are willing to spend time and money 'on enjoyable, enlightening experiences related to science and technology', and yet many people decline vaccinations, for example, or deny climate change (which we will come back to in the next chapter). Science museums involved in NISE and Ecsite are interested therefore not only in providing information, but in enhancing critical thinking around science and technology, which is to say they are using science collections in 'science engagement'. Science engagement is heir to the public understanding of science initiatives mentioned in Chapter One, with their notion of an educational deficit among the general population, and is closely akin to science *communication*, with a little more two-way dialogue thrown in. (Here I will continue my bad habit of using science as a shorthand for engagement with Science, Technology, Engineering and Mathematics – STEM.)

As well as the mysterious ferrofluid, during our search for objects in the science engagement landscape we will encounter other things on the weirder end of the spectrum: a bone screw, a Norwegian mind forest, a digital battle between collections, a museum tour for cockroaches and a swanky German coffee machine. We will confirm what we found in earlier chapters, that the best use of science collections involves telling stories about people. We will discover that engaging with science objects can be a multi-sensory experience, tangibly or digitally, for adults as well as children, despite science museums' reputation. Science objects can be *fun*.

Science Exhibitions

The first stop on our journey through science experiences is the exhibition. What sort of scale and topics are on offer, and how do objects feature alongside the other component elements of the display? Let us explore the processes behind making displays – whether permanent galleries or temporary exhibitions – and uncover how they connect objects to people.

'Nano', the 'pop-up' offered by the NISE network (see illus. 70) is a small exhibition, some 40 square metres (430 sq. ft), intended for small children and their family groups, with interpretation at different levels to prompt intergenerational dialogue. Such groups come in all shapes and sizes, so museums need to cater for different expectations: the nuclear family is only one possibility among many configurations. 'Interactive, informative, and family-friendly', it includes hands-on activities and a considerable element of play. 'Nano' can be replicated: ninety or more were on display at any one time.[5] Nanotechnology is a popular topic (similar exhibitions could be visited in Brazil and Germany, for example), with interactives seeking to illuminate and enliven the latest contemporary science.[6]

'Nano' is on the smaller, shorter and cheaper end of the spectrum of science exhibitions. Others can be giant affairs, covering hundreds of square metres, and lasting for a generation. So-called permanent galleries can be arranged chronologically, but more commonly the larger museums are arranged by theme: engineering (or 'invention'), transport, communications, computing, energy, sometimes robotics; and most science museums will have a permanent gallery devoted to space. These galleries tend to be dominated by industry or technology, as we found in the opening chapter – standing exhibits devoted to pure science are in a minority. These 'permanent' galleries often require millions of pounds or dollars of capital funding, whether philanthropic or public funds, such as the National Science Foundation support in the United States or lottery money in the UK.

The 2010s were a boom period for grand redisplays. At National Museums Scotland the entire science and technology area was reconfigured between 2012 and 2016 with the support of private donors, the Wellcome Trust, and what was then the National Lottery Fund. The Wellcome likewise supported the Science Museum's sprawling history of medicine galleries that opened in

London in November 2019 at a cost of £24 million (illus. 71). Both the Deutsches Museum and the National Museum of American History are in the process of redeveloping great swathes of their galleries over decade-long programmes, involving dozens of galleries and tens of thousands of objects; and the Polytechnic Museum in Moscow has been closed altogether for several years to change the entire display.

These large-scale overhauls tend to be undertaken stepwise, gallery by gallery. Take, for example, 'Science City: 1550–1800' at the Science Museum in London. Building on research on the history of early modern science and scientific instruments, the 650-square-metre (7,000 sq. ft) gallery, supported by funders including the Linbury Trust (one of the Sainsbury family charities that support culture), opened in late 2019. Compared to other recent redevelopments, it is unusually esoteric, exploring instrument making in the capital across 250 years. It features the greatest hits of British scientific instruments, including treasures from the George III Collection (see Chapter Two), iconic microscopes, globes and (for the first two years only) Isaac Newton's 1671 reflecting telescope. These sit among a striking setting by designer Gitta Gschwendtner. Like the Science Museum's mathematics gallery (designed by the superstar architect Zaha Hadid), 'Science City' presents scientific instruments as works of art in a calm, spacious gallery (illus. 72). Woven around them are the human stories: not only of the great and good like Isaac Newton, but of scientific instrument makers and users: not science that was 'confined to academic circles, but instead science that had practical relevance and importance to the lives of many'.[7] There are even some women featured.

'Science City' should last more than a decade, one of the long-term galleries that account for the vast majority of the floor space and visits to museums. A great deal of their attention and marketing budgets, however, is dedicated to time-limited exhibitions, designed to make a splash and lure visitors into the rest of the museum. Most are relatively small-scale, in small or medium spaces, but in science museums, as elsewhere in the sector

71 A cabinet of curiosities in the Science Museum's Medicine Galleries, opened in 2019, exploring 'humanity's experience of medicine across time and space'.

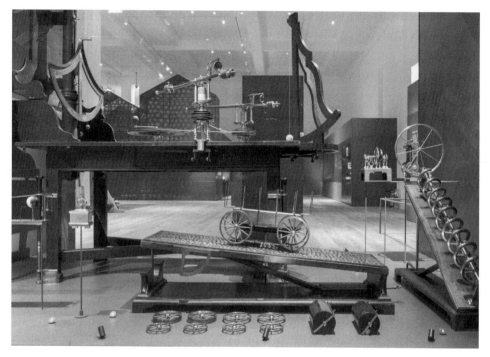

72 An object-rich case in 'Science City', 2019.

since the 1970s, these have included (in)famous 'blockbuster' exhibitions. They are heavily marketed, ripe with merchandise and other commercial opportunities and can raise a museum's profile. These do not always relate directly to science museums' own collections, gravitating rather towards crowd-pleasing topics: dinosaurs, Old Masters and, especially, ancient Egyptian mummies. For example, in 2007 'Tutankhamun and the Golden Age of the Pharaohs' was the most popular exhibit in the Franklin Institute's century-long history, attracting more than 1.3 million visits. Sometimes these are hired from specialist companies who know what spots to hit: *Star Trek* or *Body Worlds,* which tour around the world. Or they may be travelling exhibitions designed and stocked by larger museums looking to attract new audiences and geographically enhance their reach and build their brand: the Science Museum's 'Superbugs', featuring the pig cough monitor in Chapter Two, travelled to India, China, Russia and Argentina.[8] Other special exhibitions are home-grown and more particular, such as Deutsches Museum's 2019 exhibition 'Kosmos Kaffee' (Coffee World), which explored the science and technology of coffee making.[9] Neat, snappy, but too niche for a permanent display, the museum used it to promote their offerings while many of the galleries were closed for redevelopment.

'Mind Gap', in Oslo in 2011–12, was certainly a one-off (illus. 73). This strange exhibition 'aimed to examine neuroscience as practice and culture [bridging] the gap between popular culture's notion about the brain and the researcher's far more specific (lack of) knowledge about this organ'. American artist Robert Wilson helped the Norsk Teknisk Museum to design a visitor experience that mirrored the complexity and confusion of the human brain. As one might expect in a science exhibition, visitors encountered neurologists and the tools of their trade; but they did so in rooms respectively filled with mirrors, mock trees and darkness. Zoological specimens, preserved human brains and skulls, anatomical models, medical equipment and scientific instruments were 'examples of autonomous research projects being done at local places, with the groups' instruments, methods and questions . . . researchers become objects in an exhibition, suggestive of the way that the brain is scrutinized in neuroscience research'.[10] In tone and content it was markedly different from the permanent displays at the Teknisk Museum, allowing them to explore new areas in new ways, and attract new audiences: 'Mind Gap' attracted more than 250,000 visits, including 10,000 school pupils and an unprecedented volume of professional groups.

These then are a few examples of science museums' public products. But what of the processes behind them? Like other large exhibitions, 'Mind Gap' was several years in the making. Curators, educators and designers spent months talking to scientists, understanding their experiments and results, before they began the more obvious processes of selecting the objects, images and films that would go on display. They used the experimental methods that would later be codified in the LAB to explore how they might interpret the material. Wilson and his co-designer Serge von Arx worked with the team on the layout, building a model of the displays

73 The unnerving 'Mind Gap' exhibition (2011–12) at the Norsk Teknisk Museum in Oslo. The author found each section more frightening than the last.

MIND GAP

74 Modelling the 'Mind Gap' exhibition.

(illus. 74), which is common practice, and pacing out a life-size mock area (illus. 75), which is not.

In doing so, 'Mind Gap' demonstrated the qualities of a good exhibition process.[11] The team had clear and ambitious goals and intended audiences, with interpretation geared towards them. They evaluated before and during the display's lifespan and reflected carefully afterwards. And perhaps most importantly, teamwork was at the heart of the project, including close collaboration with scientists; nevertheless the development process was marked by a healthy creative tension (there was considerable negotiation around the design). Any exhibition involves not only curators, but designers, interpreters, learning staff, conservators and others. Collaboration is involved at every step: object research (finding them, acquiring them, preserving them, borrowing them, shipping them); writing the labels and any accompanying catalogue; designing the space, the graphics, the texts, the object cases and mounts; project-managing the process, the press, the marketing; cross-working with the other activities and platforms we encounter below; and of course generating sponsorship and managing large budgets.

Many of these functions are provided by freelancers, especially in smaller organizations; exhibition teams necessarily drawn from outside the host museum. Some even involve all-in co-production with different communities.[12] At any level, consultation plays an important role, with visitors, would-be visitors and external specialists providing different voices and different views. Science exhibitions rely on the expertise of scientists, medics, inventors and others: for 'Mind Gap', neurologists were closely involved in ideas and execution. Other exhibitions wisely engage with a wider range of expertise, consulting with other communities, potential audiences and even (gasp) young people. In the process of 'Science City', curators asked members of the Scientific Instrument Society to enhance the information around the collection, and worked with young members of the Worshipful Company of Scientific Instrument Makers on associated events.[13] In preparation for the exhibition 'Parasites: Battle for Survival', which we will explore below, National Museums Scotland staff spent two years working with secondary school students of the intended audience age, testing ideas, finding out what was motivating and interesting. 'There is a lot more than just the past in a museum,' remarked one participant at the end of the project; the museum team also learned a great deal in return.[14]

At the heart of the process for most exhibitions are science collections. Some, of course, like the multiple or travelling shows 'Nano' in the United States, its equivalent 'NanoAventura' in Brazil, or versions of the Science Museum Group's 'Superbugs', have no museum objects per se. But generally, at some point, most exhibition processes involve selecting a group of objects to display. Some exhibitions are object-led (in which we weave a narrative around the things we know we want to display); others are story-led (in which the objects are selected to illustrate the intended messages); most are actually a combination of the two. 'Objects were carefully selected' for 'Mind Gap', for example, 'to let visitors explore, find and think about the different ways humans through

75 The multi-skilled exhibition team pacing out the 'Mind Gap' exhibition. Curator Henrik Treimo is exiting stage left.

history have endeavoured to find answers to such questions via different approaches to the brain.'[15]

Objects are not exhibited in isolation, of course, and the bulk of the exhibition process is selecting the accompanying images, films, models, interactives and, especially, *words*. Exhibition labels and/or accompanying digital details are intended to connect the visitor to one or more of the many meanings of the objects, and writing these pithy texts is surprisingly challenging and time-consuming. A good label is poetry. In a few dozen words, the museum seeks to pack in the key chronological, geographical, functional and biographical details. It can take many hours, different people, multiple edits and testing to render the complexities of a scientific object or concept in three lines that are appropriate for the intended readership. For the benefit of those visitors that read physical or digital labels, incorporating more than just the bare technical description in a clear and elegant way can be challenging but rewarding. Take one particularly elegant label introducing a Smithsonian exhibition on the genome:

> Inside you and every
> living thing is a full
> set of instructions for
> how to grow and live . . .

> The human genome is a three-billion-part
> instruction manual written in the
> twisting, ladder-shaped molecule
> known as DNA. Despite its enormous size, your
> genome folds up so small that a copy fits
> inside every cell in your body.[16]

In this label, the visitor is first invited to 'Meet your genome', then guided from macro to micro. Elsewhere, asking questions makes for a more thoughtful experience. In 'Poker Face', an exhibit about lying, the Exploratorium label read: 'Some people look at eyes, others look at something different. How did you try to tell when your partner was lying?'[17] A label is not dumbed-down, it is exquisitely crafted.

Despite the careful craft, few visitors read every label, and different people have different learning styles – experiential, kinetic, collaborative, contemplative – so exhibit makers employ other tactics as well. Stories, messages and dialogue can be generated with interactives and working models, for example. At National Museums Scotland, visitors wind a classic static electricity generator next to a towering section of a 1950s Cockcroft–Walton generator. Elsewhere audiences can carry interactivity with them, as for example 'The Pen' at the Cooper Hewitt, Smithsonian Design Museum allowing them to interact and to 'collect' items, and then retrieve information about them after their visit.[18] As simple as a rope pulley, or as complex as a flight simulator, interactives engage several senses in the learning experience; they have the capacity to make the museum experience active rather than passive, and often involve *play*.

What, then, is the sum effect of these things, words and interactives? What stories are told by science exhibitions? Needless to say, it is difficult to generalize; there are as many narratives as there are exhibitions. What I want to focus on in the rest of this section is the surprising extent to which the science exhibition is not just about science, but also about (other elements of) culture; the best exhibitions are about not only concepts, but people. Human stories help us to understand the natural world.

The presence of authentic objects can offer a connection to the people who made or used them. What may appear a haphazard bank of instruments is the control panel of a cyclotron particle accelerator (illus. 76), on display in the Putnam Gallery at Harvard University, collected for the Collection of Historic Scientific Instruments after its last day of service in 2001, after decades of experiments and upgrades. It includes notes, a pile of cables, improvised signs, highlighter pens and, gloriously, a bolted-on pencil sharpener. Curator Sara Schechner is especially proud of the bulletin board where instructions to clean up toxic mercury and blood spills are juxtaposed with restaurant menus.[19]

Like the Manitoba II mass spectrometer (see Chapter Two), this was the site of human work, interactions, thoughts and jokes ('Remember/ I wasn't the best because/ I was the oldest/ I was the oldest/ because I was the best,' reads the printout). Science is quirky; science is messy.

Just as display-making is a collaborative enterprise, so too is the science on exhibit, in both process and product. A key element of the cyclotron and its associated ephemera is the teamwork involved in particle physics. The urge to include the standard accounts of the lone genius to attract attention and name recognition to exhibitions is quite understandable. Yes, Galileo, Darwin and Einstein will be perennially popular, but we need also to include the other members of the enterprise. Whereas clockmaker John Harrison has been valorized in the history of longitude (the 'Sobel effect', after a very popular biography), in the National Maritime Museum's 'Ships, Clocks & Stars', as in 'Science City', curators tried to balance this by including others involved in the scientific process.[20] In the latter exhibition, these included draughtsmen and the women involved in the coffee-houses of eighteenth-century London. Furthermore, as the next chapter reminds us, too often the lone genius is white.

To present different perspectives is to explore disagreement as well as teamwork. The human-ness of the processes and products of science can be engagingly represented in the museum through debates within science, and broader arguments in society about science. Museums should show that there is debate and disagreement between scientists (who are human) and within scientific communities (especially contemporary science) and that this is how science develops. They should also show that the scientific products and results can be controversial, that they do not exist in a social vacuum. And so when we opened the new science and technology galleries in the National Museum of Scotland we included important but divisive topics: Dolly the sheep representing cloning, and nuclear power in the shape of the control panel and other material from the Dounreay nuclear power plant (which had been situated as far from London as

76 The control panel of the Harvard cyclotron, on display as part of the university's Collection of Historic Scientific Instruments. Traces of the people who used it are evident – including a pencil sharpener under a clipboard on the left.

possible), which allowed us to present different perspectives on this technology.

Some science and technology museums are bolder than others at addressing controversy in their programming and exhibitions. As well as 'Mind Map', the Norsk Teknisk Museum team have a track record in provocative exhibitions, on topics such as perceptions of mental health and slave labour in civil engineering during the Nazi era. These exhibitions confront visitors with uncomfortable truths that they might not have expected in a technology museum, as does the Science and Industry Museum in Manchester in its 'Textiles Respun' gallery project on the relationship between the textile trade and slavery. Ultimately, they intend 'to tell deeper, more diverse and personal stories about Manchester's involvement in the transatlantic slave trade and crucially, to reflect how profoundly this part of that city's history continues to shape black lives today'.[21] This kind of work became the subject of intense attention in 2020 as the Black Lives Matter movement gathered momentum; but the museum sector already knew the value of inviting different voices into the room.

We will return to the relationship between science collections and race in the next chapter. Here, it is important to note that curators do not seek to use objects to provoke for the sake of it, but rather try to build confidence in their audiences so they can participate in a more informed debate. As museum consultant Elaine Heumann Gurian argues, museums should be 'safe spaces for unsafe ideas'.[22] Balancing provocation and trusted narratives is tricky, but it is worth trying.

Science Activities

We will return to the role of museum objects in debates about science in the next chapter; meanwhile, let us carry on to the next stop in our tour of science engagement activities. This intriguing and complex object, named for the 4th Earl of Orrery who commissioned it, models the movement of the planets (illus. 77). Known as a grand orrery, it is the centrepiece of the Whipple

77 The Grand Orrery made by George Adams on display at the University of Cambridge Whipple Museum of the History of Science.

Museum of the History of Science at the University of Cambridge. The Whipple, as you may remember, is embedded within an academic department, but on the day I last saw it, it was not under the gaze of dons but rather a crowd of excited schoolchildren.

Rosanna Evans was then the Whipple's learning coordinator. With 35 nine- and ten-year-olds and four assorted teachers and parents in her wake (illus. 78), she led them to a number of predetermined cases within the museum, weaving stories around the objects as she went, from the orrery at the centre of the museum to the colourful anatomical models in the upper gallery.[23] Her stories brought science to life using people: she introduced them to Charles Darwin (whom they recognized) and the astronomer William Herschel (whom they didn't). The pupils clustered around, looking at the object in question as well as darting over to other items that grabbed their attention. They asked questions, some related to the topic at hand, some not so much: 'Why are the planets named after superheroes, miss?' She used human stories to engage them not only with science but with geography, English, art and, especially, history, setting their science curriculum in its wider context. They may or may not have been listening; but they did seem to be enjoying themselves.

Their session was part of Cambridge's widening participation agenda, and after 40 minutes or so the pupils were ushered off to another part of the university. Rosanna's intention – backed up by her pedagogical expertise, specialist knowledge and energy – was to catch their attention in a busy day, 'to highlight the things I think will be the most exciting to them, most intriguing, and that will demonstrate something Cambridge University-y to them'. She knew their curriculum well, so tried 'to pick on things that they will have looked at (every child has at least one space fact) – thus things like asking them which planets are missing from the Grand Orrery etc.' And, most importantly, she is careful to 'adapt to the group'.[24]

78 A school group at the Whipple Museum of the History of Science, University of Cambridge, 2015.

The pupils' encounter with the collection is not isolated but rather an episode in a broader experience of science, formal education or otherwise. They also engage with science in other 'islands of expertise' – not only back at school, but on TV, in books, in family conversations – through which they co-construct their own meanings.[25] In this engagement ecosystem, objects are the unique selling point of museum activities. Had the class been part of one of Rosanna's longer sessions, the pupils would have had the opportunity to get hands-on with a specially selected 150-strong group of things known as the 'handling collection'. Most museums will have such a resource, sometimes including models or duplicates in place of originals. Although some curators may consider them 'sacrificial' objects, discarded cast-offs from the permanent collection with a finite shelf-life, learning officers work hard to maintain them, carefully selecting items (both original and facsimile) and supporting information to achieve their audiences' learning objectives.

Three or more sessions like this are offered per week, thus facilitating around 1,500 pupils per year; another 2,000 or so visit independently, using resources carefully prepared by the museum to support teacher- or student-guided learning.[26] And Rosanna's post (which she has since left to join another of the university's museums) was a part-time contract; scale this up for the larger organizations with many Rosannas. Overall, some 10 per cent of science museum visitors access the collections as part of organized educational groups; in the Science Museum Group's case this involves over 600,000 people per year as it seeks to maintain its position as 'the number-one UK destination for school groups'.[27]

Formal school sessions are only one of a range of activities that museums offer to bring visitors and collections together. Curators and educators go out to schools with objects or activities, such as those that comprise the 'Powering Up' programme organized by National Museums Scotland (illus. 79). As we saw in the previous chapter, collections are (sometimes) used for undergraduate and postgraduate instruction. For mixed-age audiences museums stage 'science live' and science shows such

79 Powering up: STEM engagement professional Craig Sinclair wows young people.

as 'Revolution Manchester', which greets visitors to the Science and Industry Museum with consummate ballyhoo and energetic performativity. They are hosts and active participants in festivals like the Manchester Science Festival or the Edinburgh International Science Festival. Curators and their colleagues offer themed tours, whether for specialist groups or more heterodox events like Danish art collective Superflex's 'Cockroach tours' engaging visitors with the climate crisis (illus. 80).[28] A significant proportion of science museum visitors are involved in one sort of organized activity or another. Here let us illustrate the use of objects in two more of these: drop-in object-handling sessions and the adult 'dialogue' event. Together with the Whipple session, these give a helpful glimpse of the range of activities and audiences involved in science engagement.

Some visitors arrive in a group and experience science objects in a time-bound session. Others drop in. Key elements of the distributed NISE network programme were the family 'Nano' events involving not only tubes of ferrofluid but a kit of nanotextiles, liquid crystals and paper buckyballs (illustrating the nanoscopic form of carbon found to echo the geometric architecture of

Buckminster Fuller). The hands-on activity kits distributed to 250 partner museums (together with supporting information and training resources) during annual 'NanoDays' involved tiny teacups that won't spill water, a mysterious magnetic fluid, red-coloured gold, and glass objects that seem invisible. Visitors could wander in and out as they pleased, and get hands-on with science things in a way they are rarely able to do in any exhibition.[29]

A simple but very effective type of informal activity is the 'touch table'. A facilitator – be they curator, learning professional or volunteer – arranges themselves in a suitable part of the gallery, not too busy but not too quiet. They lay out some of the handling collection on a table or trolley (at National Museums Scotland we call them 'Sparkcarts') and look as approachable as possible. Some visitors flock without hesitation, others will circle for a while. Some will pick things up immediately, others are more reticent, or will watch others interact. My biomedical curatorial colleague Sophie Goggins, for example, worked with learning colleagues to devise

80 Not your usual science activity: the 'Cockroach' tour of the Science Museum drawing participants' attention to the Climate Emergency.

81 Objects used in the 'DIY Human' Sparkcart activity at the National Museum of Scotland.

a handling box called 'DIY Human' that she offers to visitors to the National Museum of Scotland (illus. 81). It comprises seven pairs of objects, each an orthopaedic surgical instrument matched to a similar-looking household item (a bottle-brush juxtaposed with a bone brush; orthopaedic and carpentry screws). Sophie uses a gentle quiz format – 'Can you guess which is which?' – hooking those who approach the Sparkcart with familiar things then making the link to the material culture of surgery and connecting the dialogue to the museum's historic instrument collection. She has engaged with a range of audiences in this way, mostly intergenerational groups, with the adults ostensibly in a chaperoning role, but actually taking interest and encouraging dialogue with the younger members of their group. She weaves stories around the objects, of the surgeons she collected them from, of patients involved. Often visitors will volunteer their own experience of surgery. (Fortunately, Sophie has a stronger stomach than I have.)

Versions of the touch table are a common sight across museums, a simple but effective way of stimulating dialogue around objects with multigenerational visitors, commonly family groups. Despite their reputation for appealing to the young, science museums also seek to generate dialogue with activities aimed specifically at adults. These include 'science cafés' or other evening events such as 'lates', often tackling topical or difficult issues. The NISE Network ran Nanoforum dialogues, intended to tackle ethical issues around nanotech before they hit the headlines, that is, 'to encourage a "frontlash" more than a backlash'; their evaluation showed that participants increased in confidence talking about societal elements of technology.[30] The Exploratorium hosts a regular 18+ 'After Dark' series of 'lectures, demos, cinematic experiences, performances, artists [and] food tastings ... alongside 650 interactive exhibits. Topics like Sex, Cinema, Propaganda or Cannabis encourage exploration of physics, chemistry, biology or psychology.'[31] Although the use of objects varies in these activities, for our purposes here they show that science museums can be engaging, provocative places.

This informal programming is difficult to get right, however. To take one ill-fated example: in 2003 the Science Museum opened the Dana Centre, nearby but separate from the main building, specifically for dialogue events, that is, 'adult-focused, face-to-face forums ... that bring scientific and technical experts, social scientists, and policymakers into discussion with members of the public about contemporary science-based issues'.[32] They aimed at the notoriously difficult-to-reach 18–45 audience with 'X-rated science'. There were climate-themed activities, for example, and in collaboration with the British Science Association, the museum set out to attract new audiences with events three evenings a week, forty weeks a year. To this end they offered comedy, science pub quizzes, puppet shows and talk shows on topics including climate science and genetically modified foods. With high resource demand and lower take-up than anticipated, the centre closed a decade later (the building now houses more traditional offices and a library). Like other science museums in the UK and elsewhere, however, they do keep trying to engage adults with informal activities.

Digital Science

Another mode of engagement of increasing importance is the myriad of digital activity and dialogue around science collections. Take one light-hearted encounter: at 3 a.m. on 13 September 2017 an online wit posed the question during a Twitter *Ask a Curator* session: 'Who would win in a staff battle between @science-museum and @NHM_London, what exhibits/items would help you be victorious?'[33] There followed an intriguing online tournament. 'We have dinosaurs,' replied the latter, 'no contest.' And yet the Science Museum and its online acolytes suggested robots, Spitfires, poisons, mermen, fire engines and eventually a Polaris missile. Natural history combatants, during the 36-hour jaunt, included vampire fish, cockroaches, pumas, eagles, elephants and lava. Each was accompanied by a striking image and many included plugs for exhibitions and other initiatives.

This episode shows both museums and social media at their best: spontaneous, light and witty. The museum teams behind their respective feeds capitalized on their unique selling points, the objects and their characteristics, and jumped on the opportunity to promote their museums and to engage with collections. Twitter weaves around the other activities of museums. Alongside the exhibitions and interpretation explored above, social media are among a range of ways audiences can choose to interact with the collections (and, for those two days in that September, a very popular way), alongside websites, virtual gallery tours, blogs, shop portals, games and films. Of the many channels that museums use their online presence to draw attention to their collections, let us consider two in particular: first, the perhaps unprepossessing prospect of collections databases; then, we will return to social media.

Most museums now include an online public catalogue to their collections (or rather a proportion of their collections, given the cataloguing backlogs we learned about in Chapter Three). For two or three decades these have tended to be fairly standard search facilities, with a simple keyword search and a more advanced route that might include other variables such as a date or museum accession number. They are modelled on libraries, but, without the bibliographical protocols and standardization, museum catalogues have been uneven and disconnected from each other. These were helpful to the handful of specialists, but even then, if they wanted more detailed information – catalogues too often lack wider stories, context and connections – they would tend to approach the curators for more information.

More recently, science museums have been thinking harder about how to make the online connection between user and object more efficient, more friendly. Although we need additional research into how search facilities are actually used, it seems that although some will have very specific queries, most users are more casual. If these pages are 'the new museum entrances', then they should be welcoming; some museums are therefore moving away from strict search and towards 'generous interfaces'.[34] The search box model is not inevitable, it is culturally contingent; museums could, and perhaps should, also include browsing and overview facilities. 'Keyword search is ungenerous,' argued Australian digital heritage advocate Mitchell Whitelaw, 'it demands a query, discourages exploration, and withholds more than it provides.' Instead, museums should 'provide rich, navigable representations of large digital collections [which] invite exploration and support browsing'.[35] Museums tend to have far more information than is available online, and they are gradually enhancing their inconsistent catalogues by following the example of social media, seeking to join up text, image and multimedia.

The Science Museum Group's 'Search Our Collections' facility is heading in that direction at the time of writing.[36] Simple and apparently user-friendly, it encourages users to 'SEARCH > FILTER > USE'. The engine provides access to objects, photographs and archives, and to encourage browsing the search page suggests themes ('Medicine'; 'Railways'; 'Art') and highlights ('Babbage'; 'Toys & Games'; 'Psychometric Tests'). At present more than half of the collection objects have accessible records, although only 10 per cent have photographs. This number is increasing apace as the 'One Collection' project is in full flow (see Chapter Three). A select few objects such as an Enigma machine and Stephenson's *Rocket* have not only text and images, but three-dimensional renderings based on established photogrammetrical techniques.[37] Users can rotate them and access additional data about specific elements; this seems a highly appealing way of engaging with objects, and although several years into these attempts they still remain rare and somewhat primitive, there are likely to be more of them if museums can establish precisely who would use them and why.

For 3D and 2D alike, the information is travelling one-way from museum to user; this is a broadcast model rather than an opportunity for dialogue. The missing link here thus far has been a participatory interface to allow external users to enhance collections information in a credible and meaningful way. 'Transcribing, adding keywords, categorisation and more provide [not only] a deep level of engagement with the collection,' argues John Stack, digital director at the Science Museum Group, 'but also valuable contributions to enhance access to and understanding of the collection. However, the museum must be prepared to engage with the people who get involved in such projects and incorporate new perspectives back into the institution.'[38] The consultation and co-production elements are as important for digital activities as they are for the exhibitions we explored above, or even more so. What is called for is something between the strictures of peer review and the abusable freedom of Wiki crowdsourcing that will allow museum digital teams to capitalize on expertise relevant to their collections, to encourage participation without jeopardizing credibility. This is, after all, what museums have been doing for decades: taking new information and interpretation supplied by visiting researchers or remote enquirers (see Chapter Three) and adding it to the catalogue record. Online interfaces have the potential to engage users with science and to have lasting impact on the collections.

However the objects are presented, however generous the interface may be, object encounters through online catalogues are relatively low traffic – albeit qualitatively significant traffic. Social media channels have the capacity to attract rather more users. Traditionally, museums have used these channels to drive footfall to events and exhibitions. Back in October 2014, for example, in something of a marketing coup, Queen Elizabeth II sent her first tweet at the opening of 'The Information Age' exhibition at the Science Museum. It did not especially matter that she had not written it herself, but that the museum was able to link this to her previous technological adoptions: first televised broadcast, first email and so forth. Five years later she sent her first Instagram photo, again from the Science Museum, this time promoting the cryptography exhibition 'Top Secret'.[39]

Not all social media activity involves royalty. Visitors like to post pictures of museum objects and, even more so, of themselves with museum objects. The canny exhibition maker positions key selfie moments around the gallery. But we will focus here on what museums (should) do to connect with their collections. To which end, consider a striking object that is at first space-age and yet strangely familiar, even if out of context. This coffee machine (illus. 82) featured in a social media campaign around the Deutsches Museum's 'Kosmos Kaffee' special exhibition in 2019. This image is indicative of the way museums use social media to connect online audiences with technical objects: simple, shiny, effective. Most museums use channels like this that include, at the time of writing, the micro-blogging site Twitter, the social network Facebook and the image-sharing service Instagram (owned by Meta, the

82 La Cornuta ('the horned'), the coffee machine that starred in 'Kosmos Kaffee' at the Deutsches Museum, 2019–20. Posted on Deutsches Museum's Instagram feed.

parent company of Facebook). They are not yet commonly engaging with the short-form video platform TikTok. This latter medium is highly appealing to the 13–21 age bracket, and it is to younger audiences generally that museums are seeking to appeal via social media (if not that young in the case of the ageing user base of Twitter and Facebook). As well as allowing museums to be fleet of foot and response, then, social media allow them to communicate with young people and other demographics who do not physically visit.

To get a snapshot of science museums' social media activity at the end of the 2010s, I took a deep dive, gathering the five most recent posts of five museums in Europe and North America across five channels.[40] The Science Museum was the most popular of them, with 670,000 Twitter followers; others were attracting in the order of tens of thousands of users on Twitter, Facebook and Instagram, with YouTube subscribers in the low thousands. Clearly many Web surfers had not registered themselves as followers, so the overall reach will have been larger than this. Natural history is in general more popular – the UK's Natural History Museum dwarfs all the museums I looked at – but engineering, physics and astronomy are popular in visual channels.[41] Over half of all social media posts in my sample included an element of promotion, among them those promoting exhibitions like 'Kosmos Kaffee', but more were related to planned activities on site. Events also featured retrospectively as recordings, a popular practice for the Exploratorium, echoing the hugely popular practice of the Royal Institution in London. Dedicated social media activities such as 'Facebook Live' and 'ask-a-curator' elements on Twitter connect museums' expertise and objects to audiences. As well as providing a channel for voices not always heard in museums, in these fora curators are able to speak in their own voices: an important benefit, as I will discuss below.

Perhaps surprisingly, historical anniversaries were a common method of engagement: especially, in my sample, the fiftieth anniversary of the Apollo 11 Moon landing. The campaign '#onthisday' in this period also included the first iPhone, various births (including the physicist Maria Goeppert Mayer), and the first east–west transatlantic flight. Onthisday posts give museums a good excuse to showcase their own collections, whether archival images or artefacts, including, for example, National Museums Scotland marking the birth of Scottish physicist James Clerk Maxwell by posting on Instagram an image of one of his three-dimensional graphs of thermodynamic activity (illus. 83). Like the coffee machine, the model is eye-catching and unusual and invites further questions. This is when museum channels are at their best. The Canada Science and Technology Museum posts an image of one of its objects most days, the more obscure the better. Not all of them are terribly popular (an inkpot?), but the practice is a good way to provide a steady stream of object engagement, stimulating discussion and curiosity.

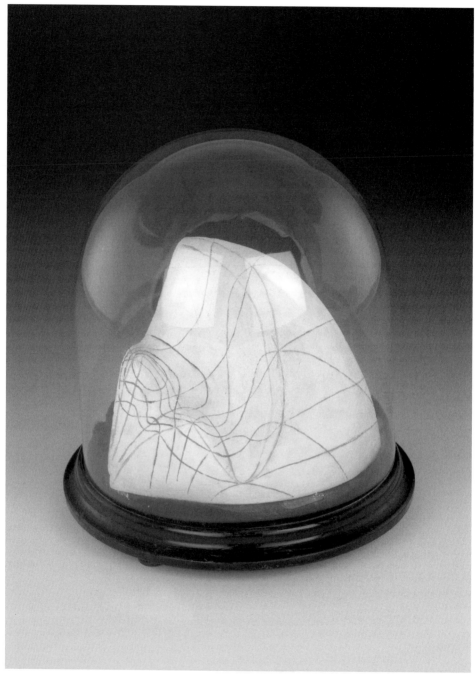

83 Thermodynamic model by James Clerk Maxwell from Josiah Gibbs's equations, Cambridge, England, c. 1875. Posted on National Museums Scotland Instagram feed, to mark the day of Maxwell's birth.

Shortly after I took my digital deep dive, the importance of virtual engagement with collections was cast into stark relief as museums shut down for a period to protect visitors and staff from COVID-19. The Exploratorium 'After Darks' are now online, for example. Some of the digital dialogue during lockdown was light-hearted, including a Twitter series 'CURATORBATTLE' sparked by Yorkshire Museum, tapping into the same appeal as the spat mentioned above. The Whipple Museum, for instance, pitched in with a rather spectacular sixteenth-century silver globe on 'fakes and forgeries' day, which allowed them to draw users into their curatorial research around forged scientific instruments.[42] Virtual walk-throughs of the galleries became more popular as people staying at home looked for different ways of engaging with culture. Onthisday tweets increased in frequency, and even though few museum staff could access collections while working remotely, they made ample use of existing object photography. More substantively, others supplied the vastly increased home-schooling demand with science engagement resources. These ranged from DIY experiments to the Royal College of Physicians of London's instructions on how to make an anatomical model inspired by their collection.[43] Museums were proactive in dispelling misinformation around coronavirus and promoting public health messages around staying at home and wearing masks, which had become deeply political.

Nevertheless the 2020 lockdown showed museums what many already knew: that social media are better suited to individuals than to institutions. Visitors do use the channels to respond, but rarely does meaningful dialogue result, and for the most part, at the time of writing, many museums are using the same strategies for these engagement channels as they are with other, slower media. Individual cultural critics and science engagers have much larger followings. In 2013 museum volunteer Emily Graslie started a YouTube channel, 'Brainscoop', about natural history collections; the Field Museum in Chicago then hired her and she hosted the channel until 2021; throughout, it bore a strong sense of her individuality and it far outstrips in popularity any of the institutions discussed here.[44] Individual science and technology curators, too, can attract large followings, respond rapidly and often use objects as hooks; but they tend to declare that their views are *not* those of their institutions. As Kirsten Riley reflected when she was social media manager at London Transport Museum, 'Our use of words is the handwriting of our personalities, so it's impossible for me not to come through.'[45] Ultimately, however, these are media of 'we' rather than 'I', and it is challenging to balance the irreverence and humour that these channels demonstrate when at their best with maintaining the authority and credibility that visitors value.

Science engagement can be done well with social media, and it can be popular: the group 'I f**king love science' has 25 million Facebook followers at the time of writing. Science museums have yet to capitalize on this. They do well when aiming for quality over quantity, and exploit simple but effective multimedia, especially posting elegant single-object images, as is the Canada Science and Technology Museum's practice. As in other areas of museum engagement, knowing where the sweet spot between what users like and what museums want to achieve is critical as they seek to connect audiences with the social and cultural elements of science, past and present, using objects as the hooks.

Science Experiences

What links these digital activities to the activities and exhibitions we explored earlier is that they are intended to connect museum collections to their audiences. Before concluding, let us flip our camera and consider the object encounter from their perspective. What sort of experience do they have?

Consider the 'Parasites: Battle for Survival' exhibition mentioned above (illus. 84). On a February day in Edinburgh, the National Museum of Scotland staged a 'Science Saturday' as part of a multi-platform programme, with the exhibition at its heart. Into the exhibition strolled a man with two young boys, seemingly a family group. Many visitors wander into an exhibition like this because they happen to be passing, or they are in the museum for a pleasant day out; this group seemed to have set out purposefully to visit the display, presumably at the adult's instigation. They spent nine minutes in the exhibition: among the longer visits on that day, but shorter than the average dwell time overall. The man was a scientist, and clearly tried to use this as an opportunity to enhance his young charges' awareness. He stopped and waxed lyrical beside the liquid-handling robot that had been used by researchers at the University of Dundee in their search for anti-malarial drugs. The boys paid little attention to the delights of automated pipetting, nor the nearby microscopes set up for visitors to see close-ups of mosquitoes. Instead they wandered off to gather 'build a parasite' stamps, the most popular element of the exhibition, which they clearly enjoyed. The trio did, however, reconvene together to play on a digital drug discovery interactive.[46]

84 Visitors engaging with 'Parasites' at the National Museum of Scotland, 2019.

This is just one of the millions of experiences generated by science collections on-site and online every day. And as digital theorist Jenny Kidd reminds us, the online/offline distinction is no longer helpful in understanding the front-end museum experience:

> starting perhaps on the [review website] Tripadvisor, moving into the What's On pages of a Museum's own website, taking in the Twitter feed, watching a documentary or reading a book, arriving in the physical museum, checking in on Facebook, listening to an audio guide, following a site map or brochure, posting their #museumselfie and maybe consulting online resources such as Wikipedia or Google as they go. For many visitors, a physical museum visit is rarely completely offline, just as an online visit is not disembodied.[47]

Blended or otherwise, it is difficult to generalize about the object experience; but I will try anyway with four broad observations.[48]

First, it is important to acknowledge that for 'Parasites' and other exhibitions, drop-in activities and museum social media channels, visitors will often wander in with no specific goals in mind. Few labels are read; few spaces prompt a prolonged stay; the average dwell time is low. They will casually browse exhibits and webpages alike, stopping when something catches the eye. Second, if there is some kind of facilitator involved, as our 'Parasites' visitors found, this can make a significant difference to the experience; and more broadly, we have seen throughout that the museum experience is shaped by people: not only (if at all) by curators but educators, facilitators, tweeters, or other members of a group or network, whether informal and multigenerational or an organized session or tour. The science museum is the setting for social interactions: not always (if at all) actually about science.[49] Third, something we do know about science museum visitors: they like to press buttons. The on-site museum visit is a multi-sensory experience, temptingly haptic, visually stimulating, noisy, conversational and, in the case of industrial history, quite smelly at times ('Kosmos Kaffee' at the Deutsches Museum had wafts of coffee rather than the usual oil and rust).

Finally, the object encounter can be emotional. Two science communicators conducted a long study of science museum experiences and concluded that visits were 'sometimes passionate, sometimes disconcerting; other times fascinating and unsettling'.[50] Artefacts are often at the heart of the emotional experience. The Boston Museum of Science's Theater of Electricity uses a historic Faraday coil and a giant 1933 Van de Graaff generator, which is nothing if not awe-inspiring. The copper cavity we met at the beginning of the book is beautiful. Nostalgia may stem from objects from a particular time – if you see your mobile phone from a particular era, for example – especially their patina of use (and abuse).[51] A visit can be visceral, especially when the human body is on display. 'Oh. My. God!' gasped one visitor as I explored the Science Museum's Medicine galleries.[52] Weaponry is not the only technology that can be frightening (I found 'Mind Gap' especially unnerving). Visitors can also feel pride, for example in national achievements: the Polytechnic Museum in Moscow celebrates Sputnik I in 1957, while the Smithsonian likes to focus on Apollo 11. (Science museums are not neutral, as we shall find in the next chapter.)

Many even find a museum visit enjoyable. We do not need to delve into the murky depths of 'edutainment' to accept that a science museum can be fun. It is not a theatre, nor an art gallery with implicit behavioural codes and a presumption of silence. For those on a formal education visit, this would tend to be more fun than the usual school day or course session; for the rest, they have generally chosen to visit a gallery or surf a site. They encounter science objects as part of a relaxed, recreational, social experience. Objects can promote genuine delight. 'What's your favourite object?' asked a friend of mine – let's call him 'P' – of his eleven-year-old, 'F', after a visit to the Henry Ford, prompting a conversation that will be familiar to many:

F: The Hot Dog Car [the 1952 'Wienermobile' (illus. 85)]
P: Why do you like it?

F: What kind of a question is that? It's a hot dog car. What else would grab my attention so much? That thing is a beauty . . . It's beautiful and touching and a hot dog.
P: 'Can you unpack that?' [P is an academic]
F: It's a *Hot Dog Car*! It's self-explanatory! You can drive it and I love it![53]

What is more, the response to collections can be as much about the building, the size of the group, the Wi-Fi, the weather, the quality of the catering and so forth. Visitors make their own routes and timetables and decide what they will focus on: they arrive at the museum or website with their own expertise, experience and motivation. They construct their own meanings from what they experience.[54]

These behaviours and experiences sometimes, but not always, overlap with what museum staff intend. Specifically, the aim of many custodians of science collections is to enhance science literacy.[55] This notion encompasses awareness of scientific concepts, confidence in using scientific terminology, and the ability to question and challenge science presented by the media and other information sources. Once again, we encounter the deeply

85 Science joy: the Hot Dog Car at the Henry Ford Museum.

political nature of activities around science collections: in many countries, governments support efforts to enhance science literacy to develop STEM skills for industrial and economic benefits. Leaving aside the pros and cons of this approach, there are problems with the concept of science literacy. It is underpinned by the assumption that science is a distinct element of life and of society, which academics working with the Science Museum Group and King's College London have tried to address by aiming instead for a broader concept of 'science capital'. This is borrowed from sociologist Pierre Bourdieu's notion of cultural capital, which (by parity with economic capital) alludes to the intangible social assets that individuals deploy within society: skills and behaviours that stem from education and position. They think of science capital (illus. 86) as

a 'holdall', or bag, containing all the science-related knowledge, attitudes, experiences and resources that you acquire through life. It includes what science you know, how you think about science (your attitudes and dispositions), who you know (e.g. if your parents are very interested in science) and what sort of everyday engagement you have with science.[56]

It does seem ambitious for museums to seek to impact upon such profound elements of visitors' experiences. How would one measure any change in these attitudes?

Indeed, how do museums measure the impact of science collections at all? Let us conclude this brief reflection on audiences by considering an overlooked function of museums, as important as any task undertaken by the staff in this chapter or throughout this book: evaluating the user experience. This is often forgotten in the sort of discussions we have about the sort of museum research we encountered in the preceding chapter. But research it is, and it should be built into any of the above endeavours from the get-go. The formative evaluation with the target audience for the National Museums Scotland's 'Parasites' exhibition we encountered above began months in advance of opening. We wanted to learn how their awareness, attitude and interest in science changed over the project, with questionnaires, discussions and group activities. Curator Sophie Goggins and learning officer Sarah Cowie asked students to stand in a line based on how much they liked science; at the beginning of the process they all stood on the negative end; by the end some of them were in the middle. This was what success looks like.

Other evaluation methods before, during and after exhibitions and activities include focus groups, tablets, observation and visitor interviews; digital teams likewise use qualitative methods and surveys and look at statistics and the level of engagement (retweets and the like). In any of these channels, the evaluator needs clear questions both for themselves and for their user groups. The NISE team, for example, wanted to know how a science display related to everyday life, and specifically how visitors used the exhibit to 'learn about the relevance of nano to their lives'; and they used video and audio recordings of the visitors talking about the content to judge this.[57] For both 'Nano' and 'Parasites', museums wanted to know, have we shifted the visitors' needle towards science? Have we enhanced their science literacy?

Part of the needle-swinging challenge is that despite what those who work in and around science may feel, in the UK at least, recent data show that public attitudes to science are already fairly positive: surveys on public attitudes to science found a majority of respondents thought science important and wanted to know more.[58] Science museums are good places to continue to learn. We know that participants in effective exhibitions and events, like the 'Powering Up' workshops at National Museums Scotland (see illus. 79), are likely to visit again, and may even be more likely to choose science at the next stage in their education. From the mountain of data from which one could draw (much of it concerned with noisiness or the lack of functional toilets), some gems also emerge. 'I think a museum needs to be both forwards and backwards looking to be engaging,' one young 'Science Saturday' visitor responded. 'I thought the museum was just about old

86 Science capital as a 'holdall'.

artefacts,' exclaimed another, 'but this is even cooler.' This we were pleased about. On the crucial issue of women in science, I was delighted when a participant in the 'Parasites' exhibition design process revealed that they now realized 'that there is not a specific gender for being a scientist'.[59]

What is not always evident is the role of objects. Researchers at the Deutsches Museum found that science objects 'possess a high degree of attraction and holding power' and they 'may trigger meaning-making-related communication' between visitors. Interestingly, it transpires that their authenticity is not as important as other elements of their objecthood, such as appearance, rarity or functionality.[60] It seems that if we are aiming to increase science literacy or broaden science capital we are succeeding. Exhibitions, events and other activities can indeed swing a visitor's needle towards science. But as we have seen, science collections stimulate a rich experience that can encompass not only changes to learning, awareness and attitudes, but much more besides.

Whether or not science objects impact upon individuals' science literacy, the experience of objects is broad, varied and multi-sensory, and often as much about joy as it is about knowledge.

Science Joy

The fun of science objects brings us back to the ferrofluid (see illus. 69). It's quirky, surprising, curious. Visitors to 'Nano' encountered it as part of a multi-sensory, sociable experience. The pupils at the Whipple were enjoying themselves, as were children and adults alike who encountered DIY Surgery on the Sparkcart or who pitched into the Twitter curator battle. Clearly science objects stimulate an engagement that isn't all about science. If, then, the *joy* of science for users of all ages is one message of this chapter, the other reiterates what we have found elsewhere in this book: that science is a cultural activity, embedded within other things we do. 'This needs to be asserted loudly and regularly,' argues the Science Museum, 'since culture is too frequently seen as synonymous with visual and performing arts, and literature . . . science must not be treated as an afterthought in cultural agendas (and vice versa).'[61] As we have seen, the best science exhibitions – the processes of which we have delved into above – are as much about people as about principles. Whether in an exhibition, event or online encounter, the science object can be a hook to explore the social, dynamic and cultural elements of science. Science is a contingent, messy and sometimes controversial activity.

This is how science museums have used objects to appeal to the visitors they want to attract and steer a course between the stultifying and the vacuous, the repellent and the superficial, to be as the fascinating and engaging. For as any science engagement professional knows, to take the abstruse elements of a technical object and render it relatable to a broad audience as part of an experience (hopefully an enjoyable one) is not a case of diluting the science. Offering meanings of science objects that might appeal to a broad range of people and motivations is worth it. Like science museum products, science museum users are varied. A few of them set out purposefully to learn more, to enhance their own science capital; far more are motivated by casual curiosity or an engaging day out. Millions of people encounter science objects in and via museums every day, many of them young, but not all; many of them have a previous interest in science, but not all. For the most part the experience is pleasant but fleeting: a glance at Instagram, a stroll through an exhibition, trying something out during an event.

The best science exhibitions, activities or tweets are those that pique users' curiosity in objects – weird and wonderful things like the ferrofluid – encouraging active learning, interaction, questioning and further enquiry rather than passively consuming something communicated to them. There are many stories to be told and meanings an object could have: what the audience makes of it is a negotiation between the visitor, facilitator (online or on-site), sometimes other visitors, and the museum. It is too easy to forget that the visitor is not a blank slate, but rather arrives at the museum or website with their own expertise, experience and motivation. The science object is therefore involved in meaning-making rather than simply channelling knowledge. The museum is a place where different people can come together around material culture and talk about science. It is what anthropologists call a 'contact zone', in which physical objects facilitate communication between different groups in a mutually trusted space.[62] As one of the NISE guys wrote in their hopes for the nano products we first encountered at the outset of this chapter, 'Science museums are conveners and community gathering places, and some are creating successful educational forums and events that bring scientists, civic leaders, policymakers, and the general public together in a neutral setting for shared learning.'[63] The savvy exhibition maker or Instagram photographer does not aim to engage a monolithic, education-orientated 'public' but rather, as Sharon Macdonald suggests, the visitor 'is positioned more democratically as a potentially knowing player, a co-worker, than a passive subject of authority'.[64] Objects like the ferrofluid channel this collaboration. They channel science joy.

87 The Bulldog HL 12 by Heinrich Lanz AG, Mannheim, 1921, in the Deutsches Museum workshop, preparing for its debut in the new permanent galleries.

5

CAMPAIGNING WITH COLLECTIONS

lumpy 1921 German tractor, manufactured by Lanz and known by its English-language nickname 'Bulldog' thanks to the unusual proportions of its cylinder head, makes a strange-looking museum object (illus. 87).[1] As an example of the first crude oil-powered tractor in the world, this 12-horsepower version used to take pride of place as a 'legend' in the Deutsches Museum's old, barn-like 'Agriculture and Food Technology' gallery. There it told the story of the impact of agriculture on everyday lives as part of an array of somewhat more conventional farming vehicles. Like 11,000 other artefacts across the galleries, it left its former position to prepare for a massive redevelopment of the museum. Five technicians tenderly guided it, refurbished and refreshed, into its new home. Despite its brand-new tyres, they pushed the 2,050-kilogram (4,520 lb) tractor on skates; the upright exhaust that makes up a good part of its 205-centimetre (80 in.) height cleared the doorframe of the gallery by millimetres.

The Bulldog helpfully encapsulates one final characteristic of science museums that has been lurking throughout the previous chapters: that despite claims to the contrary, science museums are not neutral. Between leaving the old gallery and ploughing its new furrow, the tractor was the unlikely centrepiece of an innovative and deeply political exhibition, 'Welcome to the Anthropocene'. Tackling the causes and effects of human-induced rapid climate change is one of a number of important, timely issues that science museums can use their collections and activities to address. Nothing in the previous chapters is more important than this: science museums should be political, using their rich material memory and considerable credibility to campaign for good causes. The Bulldog will take us through a discussion of how science collections can be used to address the climate crisis. Then we will park it to move on two other enduringly important issues: misinformation and human rights. As ever, other objects from other collections will pop up: a vintage camera, a prosthetic arm, some engineering tools and a humble pushbike. Each has been used to advocate change in attitude and behaviour in the audiences who encounter them.

Not everyone agrees with this approach. Ironically, in its published *Werte* (Values) the Deutsches Museum itself states that it 'is neutral, independent and committed to the principles of good scientific practice'.[2] The latter may be the case, but I do not agree with the former. Independent the museum may be: neutral it is not, even its odd little Bulldog tractor. Museums are, and always have been, deeply political entities. They reflect the circumstances of their founding (from imperial Germany to countercultural California) and have been shaped by political winds ever since (from the Victorian drive for working-class improvement to Cold War propaganda). Without clumsily assigning all museum activities as exercises of power, we can better exploit science collections if we accept that they have always been political. This non-neutrality has great potential. Museums in general, and science museums in particular, have an advantage in that they are trusted more than most other media.[3] Visitors respect the expertise of curators and material culture imbues their messages with credibility. They are 'safe places for unsafe ideas', and stakeholders on all sides value this. Science museums can be powerful advocates.[4]

There is some resistance within the sector to capitalizing on this credibility to advocate for particular causes. One science centre professional, for example, argues that 'we shouldn't be campaigning organisations ... we should keep to the principles of science'.[5] At the other end of the spectrum, others would take a far more proactive stance and use collections for all-out *activism*. The dramatic 'Klima X' exhibition, which travelled to the Norsk Teknisk Museum in 2012, for example, involved flooding the entire space with water and all visitors were given rubber boots, insisting they take action about the climate crisis.[6] Activists argue that museums should directly provoke visitors into action, prescribing what they *should do*. However, while some individual staff members are activists, it is certainly more practicable and arguably more appropriate that science collections are used instead for advocacy. This would involve sharing collections and information to stimulate dialogue and reflection on an issue, and challenging users to take action of their choosing.[7] The advocate museum identifies an issue and supports a cause – but then invites rather than demands. Ideally this does not alienate those who sit firmly in rival camps on any given issue, but rather exploits the collection as a channel, the museum as a safe space for debate.

Science museums need to walk a path between a false neutrality and the extreme of activism. Being pragmatic and canny will allow them to campaign for specific, relevant, important and timely causes. This chapter therefore unlocks something critically important: that science collections are well suited to address current, significant and deeply political issues from fake news to prejudice. We start with arguably the most profound challenge humanity faces: the speed at which our climate and environment are shifting because of human activity.

Campaigning and the Climate Emergency

Now increasingly referred to as 'the climate emergency', the rise in average temperature in recent decades is causing a rise in sea levels, extreme weather (floods in some places, desertification in others), and the accompanying economic, social and health consequences for humans, not to mention large-scale extinctions of other species. Arresting this rapid temperature change and environmental degradation requires large-scale changes in behaviour on the part of governments, businesses and individuals. Some will be influenced by the direct activism of protest groups such as Extinction Rebellion; others respond better to information, evidence and advocacy. Still others are simply apathetic on the issue. However, market research in the United States has indicated that visitors believe that cultural institutions should recommend action, and this is a key role for the advocate museum.[8] Curators can use their established trust and authority, asking questions of users in a safe space and, ultimately, influence behaviour, encouraging audiences to shift from apathy to action.

In their programming and activities, science museums in particular can exploit their place in the STEM ecosystem. To give a lively example, the Franklin Institute and the New York Hall of Science play leading roles in the Climate and Urban Systems Partnership, alongside community groups, to focus on local initiatives to combat climate change in four U.S. cities.[9] The museums act as network hubs, offering STEM activities and reaching audiences they would not otherwise attract via festivals, social media and schools sessions. Other museums also use social media to good effect in informing and campaigning around climate change. In Canada, Ingenium promotes STEM-at-home activities including a shell acidification experiment that demonstrates how climate change is prompting increased carbon dioxide and thereby affecting the oceans. Ingenium also has a visual and digital travelling exhibit, 'Climate Change Is Here', displaying striking photographs of large-scale industrial impact on the environment as well as technologies that may mitigate these risks.

Where museums come into their own is when they use their objects on display, and in activities, to evidence the causes and effects of climate change.[10] Natural history museums have a clear role to play, showing the impact of human activity on biodiversity; science and technology museums can work with and alongside them to show the human side of this equation. Material culture can take the vague and global and make it local and tangible. The materiality of the Murchison oil platform flare tip, for example, impresses upon the viewer the sheer scale of the North Sea oil and gas industry, which few experience at first hand (see Chapter Three). Its smoky patina contrasts with the clean lines of the wind turbine blades stored nearby. One mantra of Extinction Rebellion protestors, when critiquing cultural organizations in partnership with big energy, is that fossil fuel extraction should only be in museums as historic relics: this one is.

Returning again to Canada, the Science and Technology Museum uses historic and contemporary instruments in their 'From Earth to Us' gallery which tackles climate science, juxtaposing Canadian individuals' experiences of climate change with the Canadian energy industry's history:

> You can wander through a virtual mine; learn about inspiring, female miners; and discover past, current, and future mining technologies. Walk through an Energy Street where you can install a hydroelectric dam and operate a nuclear fusion reactor. Then, take a few minutes to visit the glacier – a contemplative space where you'll hear the voices of those who are experiencing climate change first-hand.[11]

Other science museums also seek to exploit interactivity to engage visitors with global warming. The Tekniska Museet in Stockholm's temporary 'Spelet om energin' (The Energy Game) took young visitors through their lifestyle choices to see the impact of energy use, building on the museum's strong history of energy collections; likewise in the National Museum of Scotland we juxtapose historic forms of energy supply with digital interactives in our 'Energise' gallery, inviting visitors to balance different energy sources, emphasizing the complex energy market and pros and cons of each.

What, then, of our trusty Lanz Bulldog tractor? It was one of the objects used in the Deutsches Museum's imaginative take on the 'Anthropocene', the new geological era prompted by human activity on the planet. '"Willkommen im Anthropozän" – Unsere

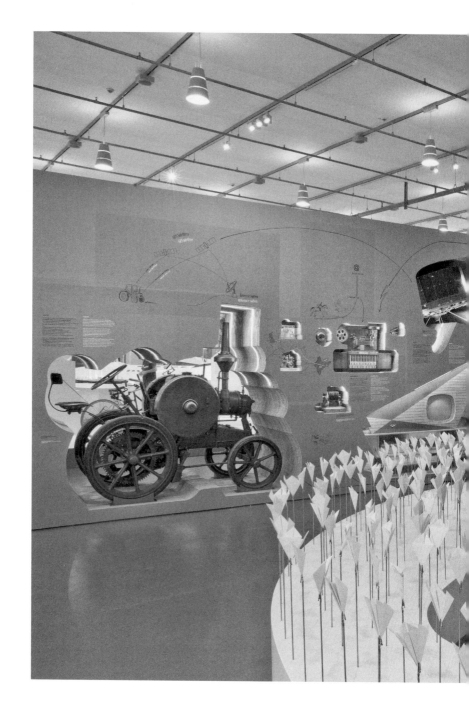

88 'Willkommen im Anthropozän' (Welcome to the Anthropocene) at the Deutsches Museum, 2014–16: artwork, digital media and historic artefact juxtaposed. The Bulldog tractor is on the left, in the foreground are the paper flowers on which visitors were invited to write.

Verantwortung für die Zukunft der Erde' (translated as 'Welcome to the Anthropocene: The Earth in Our Hands', 2014–16) took a broad, cerebral approach, inviting reflection rather than insisting on action. The tractor was at the centre of the 1,450-square-metre (15,600 sq. ft) display, acting as a hook to promote reflection on the gains and costs of industrialized agriculture over the last century. One of several 'prominent technological milestones along our path into the anthropocene', which as the harbinger of gas-guzzling tractors 'set the pace for mechanization', its brute presence was a powerful reminder of the challenges and opportunities of technology, the interrelatedness of technical systems.[12] Like the contemporary climate science instruments and other objects on display, it emphasized in specific, concrete ways how many areas of human endeavour contribute to environmental degradation. The tractor sat proudly in the 'Wall of Anthropocenic Objects' (illus. 88), where the 'lightweight, paper-based and . . . fragile structure of the wall, with its handwritten labels that allowed for easy changes, aimed to represent both the unruliness of the new machines in the garden and the openness of the Anthropocene debate.'[13] This capitalized on the particular strengths of the technology collection, drawing attention to 'the ambivalent role of technology, which contributes to many problems but also offers possible solutions, as well as humans' relationship to nature as mediated through technology'.[14]

Appropriately for its chugging centrepiece, its curators positioned 'Welcome to the Anthropocene' as 'slow media'. In the final section, visitors were invited to plant paper flowers in the exhibition, a symbolic act emphasizing their own agency. Curators then periodically harvested the flowers and the many comments written on them. 'What we should do in the future or write in the flowers is left open,' wrote curator Nina Möllers, framing the museum as advocate and the audience as active participant, 'just as in the Anthropocene itself, the visitors to the exhibition decide how they will take part in the discussion.'[15] Resisting the urge to preach, the museum was instead a place for reflection. Möllers intended it to be a contact zone between the cultural and scientific concepts of the Anthropocene, exploiting the science collection's strengths as its historic technical artefacts became boundary objects. 'In the Anthropocene,' she hoped,

> the museum cannot (and maybe no longer should) offer this assurance of certainty. Instead, the museum should [be] a forum for reflection, discussion, negotiation, and even controversy. Museums of science and technology in particular can no longer pretend to authenticate knowledge, nor can the public continue to expect this. What museums and exhibitions can accomplish and should be called upon by the public to do is to create space – literally and figuratively – for free thinking [where] visitors get the opportunity to make their own decisions.[16]

The audience was afforded the opportunity to make their own meanings and draw their own conclusions, based on solid material evidence of our impact on the planet, individually and en masse, historically and now. The Bulldog was 'chosen to carry the weight of the whole Age of Humans'.[17]

'Welcome to the Anthropocene' attracted nearly 200,000 visits during its run. The Science Museum reports that more than 5 million people have visited its climate science gallery 'Atmosphere' (illus. 89) and it hopes to attract up to 700,000 visits to its carbon capture exhibition. Science museums have reach on this topic, and this quantitative engagement is also matched by its quality. A U.S.-based study of more than 2,000 adults showed that museum visitors are more likely to know about climate change than those who do not visit; almost all visitors believe it was happening; more visitors believe it is caused by humans than those who do not visit; and compared to the 14 per cent national average, 45 per cent of frequent visitors to science museums and centres are alarmed about it.[18] What's more, they leave museums feeling they can do something about it. This, the authors of the study argue, is thanks to the trust that museums engender. With their calm authority, as museologist Robert Janes argues, 'Museums

are uniquely qualified to contribute to the issue of climate change, based on their singular combination of historical consciousness, sense of place, long-term stewardship, knowledge base, public accessibility, and unprecedented public trust.'[19]

If their reach is an opportunity, however, resource is a challenge. With limited public funding, science museums (like other cultural organizations) must look elsewhere for support, especially for programmes and exhibitions; and this topic attracts sponsors from the energy industry that many consider problematic. Whereas

89 'Atmosphere' – the 2010 permanent gallery at the Science Museum.

the British Museum and others have not renewed exhibition sponsorship from these sources after concerted pressure from groups including 'Art not Oil' and 'Culture Unstained', the Science Museum Group has continued to work with BP and Norwegian power giant Equinor (illus. 90), and staged a carbon capture and storage exhibition, 'Our Future Planet', with Shell sponsorship.[20] Despite accusations that the group is 'greenwashing' these

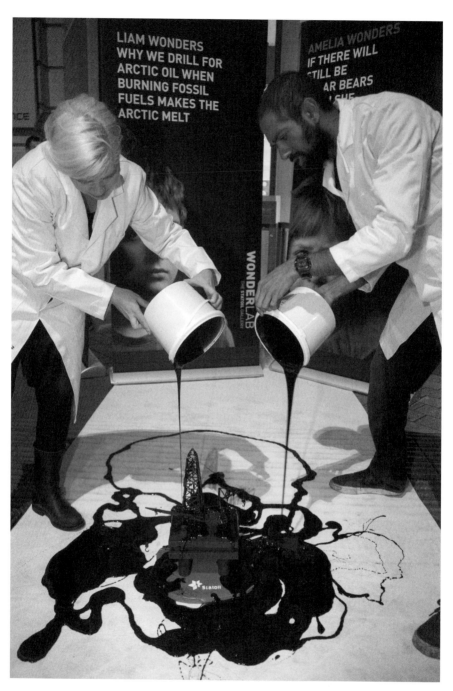

90 Protestors at the opening of the Wonderlab interactive gallery at the Science Museum (sponsored by Statoil) in 2016 pouring molasses (to represent oil) over a model rig on a white carpet (to represent the arctic).

companies' reputation, there is merit in working with them. Science collections are distinct from other museums in this respect because they collect and represent the industry and its history. As we found in Chapter Three, museums can source important objects from these organizations; and many energy companies now share museums' commitment to STEM engagement. They therefore coexist in a delicate symbiosis.

Financial support, carefully managed, can benefit both parties, even if it does open up museums to criticism. In spite of damning press around their energy company sponsors, Science Museum Group director Sir Ian Blatchford has robustly maintained these partnerships, pointing to the energy firms' support of STEM engagement and their efforts to 'find solutions to climate change'.[21] Crucially, he argues, museums retain their editorial independence, drawing parallels with media who accept energy companies' advertising.[22] Confident, authoritative museum professionals are fully capable of working with companies with clearly vested interests without ceding authorial power; and should be able to pull the plug on a relationship if it is compromising. An advocate museum can bring together both sides of a debate, working with government and corporations who might balk at working so closely with an activist organization. The Science Museum was able to attract to the launch of its 'Year of Climate Change' not only environmentalist Sir David Attenborough but the UK prime minister.[23] Advocate museums can reach places activist museums cannot.

Campaigning against Misinformation

Credible independence (not to be confused with neutrality) is also at the core of the second of the three issues I want to use to illustrate the value of non-neutral science museums as advocates: combating misinformation. Debates about the role of humans in accelerating climate change are riddled with distorted facts, but this is just one element of a wider phenomenon. Science museums are also well placed to tackle other issues, from genetic modification to anti-vaccination movements.

91 The quarter-plate Cameo camera used to make some of the so-called 'Cottingley fairy' photographs, 1917–20, which the National Science and Media Museum displayed as part of the 'Fake News' exhibition in 2017.

During the COVID-19 pandemic, for example, science communicator and Science Museum Group science director Roger Highfield wrote weekly pieces explaining the science behind the lockdown measures. His association with the museum lent more credibility, perhaps, than his two decades at the *Daily Telegraph*.[24] More broadly, providing powerful arguments around public health is important, especially online. Medical museums have used their exhibitions and virtual activities to provoke reflections around contraception, flu prevention and smoking. A good example of combating public health misinformation comes from one of the Natural History Museum's social media accounts. When botanist Sandy Knapp featured in an Instagram thread about the yew tree elements in the drug Taxol used to treat her cancer alongside chemotherapy, one user responded, 'Chemo is from mustard gas so kills more people than cancer does . . . it's awesome to see a page promote natural cures.' When misinformation promotes life-threatening behaviour, museums must be firm: 'Hi there! Just to make it clear, we're not advocating natural remedies for cancer,' responded the museum, 'We're celebrating those scientifically-backed drugs that have helped patients like Sandy.'[25]

This exchange provides a helpful illustration of a wider principle: that simply providing accurate data is not enough. However, correcting errors is not the only strategy against misinformation, and nor is it especially effective. Museums can explore the very underpinning of falsehoods, as for example in the National Science and Media Museum in Bradford's 2017 exhibit 'Fake News: The Lies behind the Truth', exploring propaganda, statistics and doctored images in the history of communications (illus. 91). Imaginatively juxtaposing past and present, 'Fake News' included manipulated photographs and social media click farms. John O'Shea, then exhibitions manager at Bradford, later reflected:

> Our role was to use our collection to give historical context, engaging visitors and exploring misinformation with them. What museums do well is taking time to unpack concepts, breaking down complex ideas, raising questions rather than providing answers. By gathering examples from the past, we highlighted how different the phenomenon is now, for example in the speed at which misinformation circulates.[26]

Using science collections' unique selling points – past and present, material and visual – the museum took a firm stance, but ultimately let users decide for themselves. They unpacked the processes behind the science and technology, asking questions and engaging in dialogue.

Curators cannot fight lies with more and more facts: outright contradiction is often perceived as condescension at best, or at worst an outright threat to one's world view.[27] The best way to counter misinformation (false information shared without necessary malice) and disinformation (lies spread intentionally to cause harm) is to promote scientific literacy, to enable users to judge the quality of information for themselves. Advocate museums are able to bring in objects, experts and expertise as part of human stories, in a safe place, to let consumers make up their own minds and select their own courses of action. In the last chapter we explored the ways museums seek to enhance our audiences' relationship with science, whether that is considered 'science literacy', 'science capital' or 'science joy'. However conceived, the benefit is clear: sparking curiosity, and offering tools of discernment, can help people choose to make better decisions. The trust, thought-provocation and even the fun we saw in the last chapter can have a life-saving function. Museums can help visitors to be active, engaged and discerning citizens.

Campaigning for Human Rights

Finally, and perhaps surprisingly, science museums have the opportunity to catch up with other museums and to advocate for human rights.[28] Among the many social iniquities that cultural organizations address, science collections are especially relevant in addressing prejudices associated with disability, and with ethnicity.

Katherine Ott, a curator at the Smithsonian, has for many years used collecting, exhibitions and digital engagement to campaign against racism and ableism ('the belief that people with disabilities are inferior to the able-bodied').[29] As part of one of the many advocacy projects she has been involved in over her career, she collected the bicycle that had belonged to Junius Wilson, a deaf African American man confined to a psychiatric facility in

North Carolina for seventy years (illus. 92). In his youth Wilson communicated via a form of sign language taught only to African Americans, which no one else understood in the legal trial he endured after he was wrongfully arrested. Despite having no psychiatric condition, he was incarcerated in the 'State Hospital for the Negro Insane' and sterilized. His bicycle's presence in the national collection speaks of the limited freedom he gained later in life. This mundane artefact represents Wilson's agency; its presence is a subtle element of the ongoing campaign for disability rights.

Katherine is not alone. At the Science History Institute's museum in Philadelphia, curators set out to collect artefacts in association with an oral history project exploring the lived experience of scientists with disabilities. The institute set out to address the impact of disability rights on scientific professions; its manifestation in material culture encouraged reflection on the role of scientists with disabilities 'as active knowledge producers and not

92 Junius Wilson's Schwinn safety bicycle on display in the Artifact Wall of 'The Americans with Disabilities Act, 1990–2015' at the National Museum of American History, 2015.

simply . . . as benefactors of technological developments, often in the form of medical devices and treatments'.[30] Their challenge, shared by many colleagues collecting contemporary science, was that many of the participants' instruments were no different from other scientists', and furthermore they were still in use. One creative solution was to display a periodic table without the 22 elements discovered by scientists with disabilities (illus. 93).

At the National Museum of Scotland we also use artefacts to try to address the thorny relationship between disability and technology. We have an extensive collection of prosthetic limbs, including the award-winning Edinburgh Modular Arm System, known as 'the world's first bionic arm' (illus. 94). We collect not only material from the engineers and medics who developed these devices, but testimony from those who used the early

93 'Science and Disability' at the Science History Institute, Philadelphia, 2019. The periodic table shows which elements were discovered by scientists with disabilities.

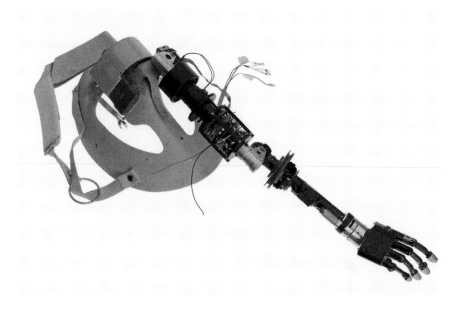

94 The prosthetic Edinburgh Modular Arm System, first used by hotelier
Campbell Aird, now on display in the National Museum of Scotland.

prototypes. This is an important perspective. Hotelier Campbell
Aird, who had lost his arm to cancer, was pleased to use the first
Edinburgh arm. 'It will enable me to do simple things like tie my
own shoelaces,' he said. But not all users have such positive experiences. Allan and Yvonne are among the survivors of those born
around 1960 with life-impacting birth defects after their mothers
were prescribed the drug Thalidomide during pregnancy. As
young children they were assigned innovative prosthetic limbs,
but ultimately neither opted to use them. 'Approaching teenage
years, the arms made me "look normal" but were heavy and cumbersome for my small body frame,' Yvonne told curators; 'I needed
help with dressing when wearing them, but without them I could
manage independently.'[31]

This testimony of non-use accompanying these technologies
is a powerful and important balance to accounts of superhuman
valour and ground-breaking firsts. We aim to balance perspectives between the medical model of disability (that disabilities
are problems to be fixed) with a social approach (that disability

is part of human life, and society should change). We try in our interpretation to stimulate awareness and dialogue in our visitors. By taking an advocacy approach we were able to collaborate effectively not only with users, but with surgeons and prosthetists, who might have been alienated by an activist stance.

Junius Wilson's bike also reminds us of another field of human rights that science museums engage in: efforts to combat racism. Here, however, it will take some time to ensure that museums are heroes rather than villains, for despite curators' best efforts to be anti-racist advocates in recent decades, there are ways in which science collections are imbued with the very prejudices we wish to combat. And science collections have some catching up to do within the museum sector. The apparently innocuous objects seen here are some of the 8,430 objects from the workshop where engineer James Watt worked in the attic of his house in Birmingham, in the years around 1800 (illus. 95). These iconic items were 'saved for the nation' by the Science Museum in 1924 and are on display in the Energy Hall there. And yet we know that this giant of the Industrial Revolution – Watt was instrumental in developing the steam engine – was closely involved in the transatlantic slave trade. We have known for some time that the Watt family and the Boulton and Watt company were engaged in commerce and trade with the West Indies that benefited from slave labour; in-depth research recently revealed that Watt himself trafficked a Black child, Frederick, in 1762.[32]

Museums' ongoing efforts to address histories of colonialism, race and racism were thrust into the limelight after the death of George Floyd in the United States in May 2020. The resurgence of Black Lives Matter, a movement founded in 2013 to protest against violence against Black communities, pushed public organizations, including science museums, to address their own positions. Social media allowed a fleet-of-foot response. The Exploratorium, for example, swiftly announced its opposition to 'systemic racism and oppressive violence'; Ingenium likewise followed suit with the simple message, 'no racism is acceptable.'[33] The Science Museum, despite reports of staff agitation, took

95 James Watt's garret workshop at the Science Museum – seemingly innocuous objects whose links to the transatlantic slave trade the museum is now exploring.

somewhat longer to explicitly 'stand against racism', but Sir Ian wanted to substantiate their public statement with specific actions 'with our collection and the stories we tell'.[34] Crucially, then, this involved not only political commitment but indications as to how they might involve their objects and practices. 'The choices we make about what to research', reflected chief curator Tilly Blyth, 'can help us to understand ... the new roles the collection might play in creating spaces that are open for everyone.'[35]

Blyth also pointed to 'the role objects in our collection had in supporting colonial structures'.[36] Museums not only reflect the inherently inequitable structures of contemporary society, but they helped to establish the very hierarchies upon which white privilege is based. The events of summer 2020 did not suddenly provoke this realization: museums have been wrestling with these issues for decades. I experienced my own steep learning curve some years ago when I worked on an unsettling project exploring the relationship between collections and race. We found that museums were complicit in the construction of physical and cultural hierarchies that underpinned racist thought from the Enlightenment until well into the twentieth century, in ironic contrast to the inclusionary role they now seek to play.[37] Much of the attention in this area has been around the anthropology collections that housed the human remains used to develop racial hierarchies. However, other types of collections also contributed to the construction of ethnic difference. Natural history museums classified human races in the scale of animal life placing European men at the pinnacle; science and technology museums collected material to emphasize the technological supremacy of Western culture. As we saw in Chapter One, the great bulk of collections arrived during the decades around 1900 – the 'payday of empire' – and that mindset is reflected in them.[38] Museums, as engines of difference, helped to build the very concept of race.

Still today, inequalities are evident in the staffing of museums: all the organizations I explore in detail in this volume are dominated by middle-class white people. The Exploratorium addressed this head-on by committing to

> look within, individually and institutionally, and acknowledge the ways in which we are a part of the problem. We acknowledge that the Exploratorium, like too many museums, historically and currently, has been predominantly led and visited by white people. We acknowledge that we haven't done enough to hear and highlight Black voices both internally and externally.[39]

This is challenging, internally and externally. Diversifying the museum workforce will involve addressing the economic disparities that allow wealthy would-be curators to volunteer unpaid in order to increase their chances of employment. And encounters between museum professionals and external individuals, particularly those from diaspora communities, still bear traces of colonizer meeting colonized.

Museums were not neutral in their founding and development, and they must not pretend to be neutral now. Science museums are slowly following other museums in addressing their own colonial and racist histories and current approaches to ethnicity in many ways, three of which I want to address here. First, science museums are being more open about their collections' historic links with slavery. Take Watt's relationship with the transatlantic slave trade, which is now (gradually, marginally) more evident in on-gallery interpretation in the Science Museum's sites, the National Museum of Scotland and elsewhere. The Science and Industry Museum in Manchester was already rendering the links between slavery and cotton explicit in its 2018 project 'Textiles Respun': they are also addressing these links in their public programming around those displays. However, many science museums have been inactive to date on this issue, leaving social history and other collections to take the lead; they should now use their collections to reveal how embedded enslavement was in the Industrial Revolution and in the modern world.

Second, science collections can be used to address the science around race, ethnicity and biology. There have been exhibitions devoted to this for some time now, including the Minnesota Museum of Science's 'RACE: Are We So Different?' (whose approach is evident in the title) and the Ontario Science Centre's 'A Question of Truth', which addresses 'a belief in European, white male superiority [that] resulted in the use of science as a tool for prejudice, discrimination and atrocities'.[40] These exhibitions are powerful advocates, prompting visitors to question the neutrality of science and their own attitudes and beliefs. The latter gave one visitor pause to reflect that 'certain people were considered inferior [for example] black people that were given syphilis to research the causes and effect in terms of disease', which she thought 'was horrible to see'.[41]

Let us pause here to consider the Norsk Teknisk Museum's 2018 exhibition 'FOLK – From Racial Types to DNA Sequences' (illus. 96), which juxtaposed the history of race science, including phrenology and eugenics, with contemporary work on population genetics in order to challenge contemporary racist attitudes.[42] The exhibition reinterpreted troubling historic objects including anthropological slides, scales for measuring human hair and eyes, waxwork anthropometric models and human skull casts. They focused especially on the study of the Scandinavian indigenous Sámi people. The curators worked with members of the Árran Lule Sámi Centre to identify individuals who were the subject of early twentieth-century anthropological study; even if the museum could not give their voices back to these objectified individuals, they at least were given their names. Importantly, this historic work prompted visitors to reflect on their own ethnic identities, their implicit and explicit biases. 'FOLK' was intended to explore 'the capacity of museums to stimulate critical reflection and dialog on constructions of human difference, and thereby to serve as agents of social change'.[43] In exposing the history of difference-making in science, they advocated respect for both difference and same-ness.

Finally, science museums can do much more to address the lack of diversity in their own collections, displays and programmes.

96 'FOLK – From Racial Types to DNA Sequences' (2018) at the Norsk Teknisk Museum, Oslo.

The Exploratorium committed to 'do more to highlight Black scientists, scholars, artists, and community members in programs and exhibitions'.[44] This can involve simple online activities such as commemorating the birthday of Marie Van Brittan Brown, the African American inventor of the CCTV home security system, or drawing attention to people of non-European descent represented in collections.[45] But there remains a long way to go. Take for example the NASA mathematician Katherine Johnson, who was the central character in the book and film *Hidden Figures*.[46] The National Museum of African American History and Culture has a portrait of her, but a catalogue search reveals little else to represent her in the national collection (she passed away in 2020, and at the time of writing the fate of any material estate is unclear). Johnson's apparent under-representation is illustrative of the challenge of representing hidden figures in science museums dominated by stories of individual white men (see Chapter Five). Using images, records and digital evidence we should be expanding our collections and activities to show the role of people of colour (and other deleted groups) in science.

Science museums should address the histories of enslaved people, racism and diversity through and with their collections, promoting an understanding of the history and legacies of difference and advocating a more equitable society. They should use objects like the Lanz Bulldog and Junius Wilson's bicycle to address the climate emergency, challenge misinformation and promote human rights, moving from complicity in inequalities to combating them. 'As lifelong learners, educators, scientists, and artists', commits the Exploratorium, 'we must challenge racial inequities and injustice and work towards a world that respects, protects, and celebrates all its people.'[47] There is a long way to go, and this is not easy for staff or visitors. When museums use collections to challenge prejudices, or to reveal uncomfortable truths about the state of the planet and the humans living on it, they engender thoughtful responses. As one visitor observed of 'A Question of Truth',

I am a young black female and it is about time someone openly came out and cleared up some of these issues. This exhibit is definitely a wake-up call to all youths and its striking realism and honesty about some of the past and present issues will help to put an end to so many stereotypes.[48]

Long may such advocacy efforts continue.

6

LIVELY COLLECTIONS

Since we set out together to explore their contents, their history, their hidden places, their staff, their use and their users, what have we learned about science collections? We found that science museums are not always about science, but also about industry, agriculture and transport; about health and medicine, and – especially – about society. Even though we focused on material things, we found that objects in collections are vastly outnumbered by texts, images and digital files. Despite this quantity and diversity of material, we found that far from being encyclopaedic, throughout their history and in their collecting practices, science collections have been surprisingly idiosyncratic. We found therefore not only considerable strengths, but weakness, and that they excrete as well as digest. We found that the great majority of the collections are not on display, but rather in vast collection facilities, where they are preserved, used and studied by conservators, researchers and others; and we found that science collections appeal to those of all ages, and can evoke meaningful emotional connections with these audiences. These objects can provoke fear, hope, revulsion and even joy. Finally, we found that by telling stories with objects, science museums can make a difference, provoking reflection and action to improve humanity's lot.

If these have been the results of our specific enquiries of collections in the chapters above, let us conclude our journey by reflecting on them as a whole. Specifically, as we have found again and again, science museums are as much about society as they are about science, as much about culture as they are about the study of nature. Embedded in their collections are stories of people.

Science Is Human, Too

Think back to the Mignon typewriter in Chapter Three. Together with its associated images and penumbra of documentation, it told us about the history of typewriters: but the object also had its own specific biography, from Great War Germany to radical Edinburgh. Technologies embody human activity, and through their individual journeys to the collection, and the people they encountered on the way, they each have a particular human story to tell. Science collections thereby have a rich cluster of social narratives running through them. As assemblages they are subject to the contingencies and passions of individual humans, from the revolutionary French cleric Henri Grégoire to the conservator Sarah Gerrish. This is especially clear in the case of Founding Fathers (and they were nearly all men) from the Exploratorium's Frank Oppenheimer to Oskar von Millar of the Deutsches Museum. Their obsessions (and egos) remain evident to this day in the institutions they founded, overlaid with the passions and peccadilloes of generations of curators and others involved in the collections since then.

This complex human-object tangle is also true of other collections. It is one of the many parallels and commonalities we have found between science collections and natural history, social history, art and other museums. They have common pedigrees and similar collecting practices, and they share storage facilities and exhibition processes. Science objects like the CERN copper cavity have sculptural aesthetics, like a pot in the archaeological collection or a statue. Science is undeniably a unique form of human activity – an especially well-resourced and impactful activity – but it is a human activity. Collecting the history of science is akin to collecting social history because the history of science *is* social history. 'Science is embedded in the world around it,' argues Tilly Blyth of the Science Museum Group, 'and only by seeing science as part of our broader culture can we begin to understand and interpret it in the rich and multifaceted way it deserves.'[1] Science collections play an important role in helping us to understand the relationship between science and (the rest of) society, as one curator reflected in relation to the 'Question of Truth' exhibit at the Ontario Science Centre: 'I think the role of science museums is to get people to be aware that science is embedded in their culture. It's part of their everyday life. And they have to have some agency in what's going on.'[2]

Well-interpreted science collections are about people as well as concepts, and, most of all, they involve stories. The skill in telling such stories is to render science relevant, meaningful and personal, as we saw with the Harvard cyclotron control deck with its takeaway menus. These stories are rooted, embedded and woven in and around the objects in collections. The Science Museum's bicycle from CERN tells us about the scale of particle physics; Junius Wilson's bicycle tells us about the recovered agency of a disenfranchised individual. These stories need not be of lone geniuses making groundbreaking discoveries (although many of them still are). Objects are revealing of the teamwork involved, as well as the messiness, the serendipity and the improvisation of science. Technological dead ends and unrealized prototypes challenge the image of indefatigable and incontrovertible scientific progress; rather, failure and foible are human stories of science in all their glorious ambiguity. From protests against nuclear energy, or cloning, to the links between cotton-processing technology and enslavement, objects speak of conflict and controversy. The tales science collections tell are not always comfortable to hear.

Science Collections Are Alive

The human stories wrapped around the objects in science collections accrue and develop; some are lost and forgotten, others re-emerge. When we explored museum stores in Chapter Four we found that even these warehouses of science are dynamic places. We found that objects, images, texts and data arrive and depart on loan and from display and the workshop. The Bulldog tractor is pushed into its new home; other objects trundle back and forth on mobile shelving; digital entities are emailed to and fro. Even though the vast majority of them are away from the public eye, apparently in repose, we have seen that, like a stop-motion coral reef, science collections over the decades are far more dynamic than one might think. The exhibitions and activities these collections enable, from Nanodays to tweets, stimulate new knowledge and understandings. Science and technology collections may not be as alive as those in zoos, but they are certainly lively.[3]

At every step we have seen the driving force behind science collections' liveliness: the personnel involved. As well as being about people, science collections are of course staffed by them. In my experience these professionals and amateurs are united by their passion for the collections in their care – their 'object-love'. Fuelled by this passion, science collection staff need to be multiskilled and fleet of foot. Curators and their colleagues need to be adept at research, collecting and engagement – and many more skills besides. For those organizations lucky enough to have resources for separate specialist posts, the conservators, enablers and curators need to be able to work with each other. Science exhibitions and other processes, such as Oslo's exploration of neuroscience in 'Mind Gap', rely on effective teamwork.

Science collections are kept alive not only by these internal collaborations but by effective partnerships, and via the thriving networks between science collections, other kinds of museums, and other elements of the STEM ecosystem. Museums collect things together to share (illus. 97); they transfer and redistribute items to each other. Increasingly they will interlink their collection

97 The control room of the Dounreay nuclear power station, originally co-acquired by National Museums Scotland and the Science Museum Group in 2015.

search facilities to enable users to access different collections in a single click. They connect with laboratories and universities to exchange material culture and expertise; they share stories with science centres and YouTube channels such as *IFLScience*. They would do well to follow the group of u.s. natural history museums who made a 'declaration of interdependence'.[4]

Regardless of their type, at the heart of this declaration is the insistence that museums are connected and also relevant.[5] One of the ways they remain so, as we have seen throughout this book, is to collect and engage not only with science past, but with its present. Science collections have always swung like Foucault's pendulum between then and now; cutting-edge technology already rubbed shoulders with scientific relics in the early years of the Musée des arts et métiers two centuries ago. This has always

98 Science joy: people enjoying the interactive 'energy wheel' when it was part of the displays at the National Museum of Scotland.

brought with it a tension: how can curators engage universal ahistorical scientific principles with artefacts that are chronologically specific? The answer lies in treating contemporary science just as we do historic practice – as (wonderfully) contingent and human. There is a bewildering profusion of contemporary material from which to choose, but there always has been and always will be: so we select important, interesting and relevant science 'stuff'. These new things may be digital or material, textual or visual. Remember the pathological pigs in Chapter Three? The Science Museum collected the microphone to listen to them coughing and the software to detect the poorly porcines. Collecting the multimedia present, now and in the future, will keep these collections – like the pigs, hopefully – alive.

Finally, and I hope obviously, we keep science collections alive for the most important people involved with them – their visitors, audiences and users. Chapters Five and Six framed the material culture of science as boundary objects and the science museum as a contact zone between different communities. Scientists, technologists and medics ascribe particular values to science collections; museum professionals do their best to render them engaging for on-site, online and behind-the-scenes visitors, who each in turn make their own meanings. These users may be motivated by specific curiosity, by browsing social media or simply by a nice day out; as many encounters with science objects are about leisure as about learning. We found that despite their reputation, these audiences include adults as well as children.

The lively character of science collections is especially evident in these interactions with users of any age – and these interactions are multi-sensory and often emotive (illus. 98). Hands-on experiences are not limited to push-button interactive galleries, but can also be found in learning events and the collections facility. Encountering a copper cavity, Foucault's pendulum, the OncoMice, a typewriter, a ferrofluid sample, the Bulldog tractor or any other object during a museum visit can be noisy, bustling, conversational and visually delightful. It can be unsettling or even frightening; but so many emotional engagements with objects whether on gallery, in store or online, are simply joyous. I hope *Curious Devices and Mighty Machines* has sparked your curiosity.

REFERENCES

Introduction

1 Nick Richardson, 'At the Science Museum', *London Review of Books*, 6 March 2014, p. 25.

2 John Durant, 'Science Museums, or Just Museums of Science', in *Exploring Science in Museums*, ed. Susan M. Pearce (London, 1996), pp. 148–61 [p. 152].

3 See, for example, John Edwards, 'Big and Oily: Collecting the North Sea Gas and Oil Industry', *Social History in Museums*, xxxi (2006), pp. 27–32; T. N. Clarke, A. D. Morrison-Low and A.D.C. Simpson, *Brass and Glass: Scientific Instrument Making Workshops in Scotland* (Edinburgh, 1989).

4 See Samuel J.M.M. Alberti and Elizabeth Hallam, eds, *Medical Museums: Past, Present, Future* (London, 2013); Samuel J.M.M. Alberti, *Morbid Curiosities: Medical Museums in Nineteenth-Century Britain* (Oxford, 2011).

5 Ecsite, 'Norsk Teknisk Museum', www.ecsite.eu/members, accessed 19 December 2021.

6 See, for example, Samuel J.M.M. Alberti, 'Constructing Nature behind Glass', *Museum and Society*, vi (2008), pp. 73–97; Eric Dorfman, ed., *The Future of Natural History Museums* (Abingdon, 2018).

7 The International Council of Museums has more than 30,000 members in 137 countries: International Council of Museums, 'Museums Have No Borders', https://icom.museum/en, accessed 3 July 2021.

8 Dagmar Schäfer and Jia-Ou Song, 'Interpreting the Collection and Display of Contemporary Science in Chinese Museums as a Reflection of Science in Society', in *Challenging Collections: Approaches to the Heritage of Recent Science and Technology*, ed. Alison Boyle and Johannes-Geert Hagmann (Washington, DC, 2017), pp. 88–102.

9 Marianne Achiam and Jan Sølberg, 'Nine Meta-Functions for Science Museums and Science Centres', *Museum Management and Curatorship*, xxxii (2017), pp. 123–43.

10 International Council of Museums (2007) definition: https://icom. museum/en/resources/standards-guidelines/museum-definition, January 2020. At the time of writing, ICOM has spent five years discussing a new definition without reaching consensus.
11 National Museums Scotland, *Shaping the Future: Strategic Plan 2016–20* (Edinburgh, 2016), p. 7.
12 Science Museum Group, *Inspiring Futures: Strategic Priorities 2017–2030* (London, 2017), p. 12; emphasis added.
13 Museo Nazionale della Scienza e della Tecnologia Leonardo da Vinci, 'Mission', www.museoscienza.org, accessed 19 December 2021; National Museum of American History, *Strategic Plan 2013–2018* (Washington, DC, 2013), p. 9.
14 Marta C. Lourenço and Lydia Wilson, 'Scientific Heritage: Reflections on Its Nature and New Approaches to Preservation, Study and Access', *Studies in History and Philosophy of Science*, XLIV (2013), pp. 744–53 [p. 752].
15 Ingenium – Canada's Museums of Science and Innovation, *Summary Corporate Plan 2019–2020 to 2023–2024* (Ottawa, 2019), p. 5.
16 Polytechnic Museum, 'Mission', https://polymus.ru/eng, accessed 19 December 2021.
17 Deutsches Museum, 'Mission', www.deutsches-museum.de, accessed 19 December 2021.
18 Science Museum Group, *Inspiring Futures*, p. 11.
19 Ecsite, 'Norsk Teknisk Museum'.
20 Rijksmuseum Boerhaave, 'About Us', https://rijksmuseumboer-haave.nl/engels, accessed 19 December 2021.
21 Simon Schaffer, 'Object Lessons', in *Museums of Modern Science*, ed. Svante Lindqvist (Canton, MA, 2000), pp. 61–76.
22 Sandra H. Dudley, 'Museum Materialities: Objects, Sense and Feeling', in *Museum Materialities: Objects, Engagements, Interpretations*, ed. Sandra H. Dudley (Abingdon, 2010), pp. 1–17 [p. 6]; original emphasis.
23 Jack Challoner, *Science Museum: The Souvenir Book* (London, 2016); Volker Koesling and Florian Schülke, *Man, Technology! A Journey of Discovery through the Cultural History of Technology* (Berlin, 2013); David Souden, ed., *Scotland to the World: Treasures from the National Museum of Scotland* (Edinburgh, 2016).
24 Alison Boyle, 'Stories and Silences in Modern Physics Collections: An Object Biography Approach', PhD thesis, University College London, 2020, p. 258; see also Alison Boyle and Harry Cliff, 'Curating the Collider: Using Place to Engage Museum Visitors with Particle Physics', *Science Museum Group Journal*, 2 (October 2014).
25 Garth Wilson, 'Designing Meaning: Streamlining, National Identity and the Case of Locomotive CN6400', *Journal of Design History*, XXI (2008), pp. 237–57 [p. 254].

26 On colours of science and medicine collections, see Shaz Hussain, '50 Shades of Beige', https://blog.sciencemuseum.org.uk, 11 April 2018; David Pantalony, 'The Colour of Medicine', *Canadian Medical Asssociation Journal*, CLXXXI (2009), pp. 402–3.

27 Koesling and Schülke, *Man, Technology!*

28 Constanze Hampp and Stephan Schwan, 'The Role of Authentic Objects in Museums of the History of Science and Technology: Findings from a Visitor Study', *International Journal of Science Education*, Part B, V (2015), pp. 161–81.

29 Rachel Sharp, '20/20 Vision', *Scottish Review*, 16 October 2017.

30 Linda Sandino, 'A Curatocracy: Who and What Is a V&A Curator?', in *Museums and Biographies: Stories, Objects, Identities*, ed. Kate Hill (Woodbridge, 2012), pp. 87–99.

31 Hilary Geoghegan and Alison Hess, 'Object-Love at the Science Museum: Cultural Geographies of Museum Storerooms', *Cultural Geographies*, XXII (2015), pp. 445–65 [p. 459]; original emphasis.

32 Rowan Moore, 'Wonderlab: The Statoil Gallery', *The Observer*, 9 October 2016.

33 Ipsos MORI, *Wellcome Trust Monitor, Wave 3* (London, 2016).

34 John Durant, conversation with the author, Cambridge, MA, 23 April 2018.

35 Association of Science and Technology Centers, *Science Center Statistics* (Washington, DC, 2017). See also Fiammetta Rocco, 'Temples of Delight', *The Economist*, 21 December 2013; David Chittenden, 'Roles, Opportunities, and Challenges – Science Museums Engaging the Public in Emerging Science and Technology', *Journal of Nanoparticle Research*, XIII (2011), pp. 1549–56.

36 Ipsos MORI, *Wellcome Trust Monitor*.

37 Emily Dawson, *Equity, Exclusion and Everyday Science Learning: The Experiences of Minoritised Groups* (London, 2019).

38 John Durant, conversation with the author, Cambridge, MA, 23 April 2018.

39 Steven Conn, *Do Museums Still Need Objects?* (Philadelphia, PA, 2010).

40 Joshua P. Gutwill and Sue Allen, *Group Inquiry at Science Museum Exhibits: Getting Visitors to Ask Juicy Questions* (San Francisco, CA, 2017).

41 Richard Fortey, *Dry Store Room No. 1: The Secret Life of the Natural History Museum* (London, 2008). Others I have enjoyed include Lance Grande, *Curators: Behind the Scenes of Natural History Museums* (Chicago, IL, 2017); Steven Lubar, *Inside the Lost Museum: Curating, Past and Present* (Cambridge, MA, 2017); and Nicholas Thomas, *The Return of Curiosity: What Museums Are Good for in the 21st Century* (London, 2016).

42 Three of my favourites are from a little while ago: Victor J. Danilov, *Science and Technology Centres* (Cambridge, MA, 1982); Stella Butler, *Science and Technology Museums* (Leicester, 1992);

and Sharon Macdonald, *Behind the Scenes at the Science Museum* (London, 2002). More – and more recent – volumes can be found in the Select Bibliography.

1 How Collections Came to Be

1 The literature on the history of science museums is sizeable. For helpful surveys, see Alan J. Friedman, 'The Extraordinary Growth of the Science-Technology Museum', *Curator: The Museum Journal*, L (2007), pp. 63–75; Oliver Impey and Arthur MacGregor, eds, *The Origin of Museums: The Cabinet of Curiosities in Sixteenth- and Seventeenth-Century Europe* (Oxford, 1985); Silke Ackermann, Richard L. Kremer and Mara Miniati, eds, *Scientific Instruments on Display* (Leiden, 2014); Elena Canadelli, Marco Beretta and Laura Ronzon, eds, *Behind the Exhibit: Displaying Science and Technology at World's Fairs and Museums in the Twentieth Century* (Washington, DC, 2019).
2 Fiona Candlin, *Micromuseology: An Analysis of Small Independent Museums* (New York, 2015).
3 Paula Findlen, *Possessing Nature: Museums, Collecting, and Scientific Culture in Early Modern Italy* (Berkeley, CA, 1996).
4 See, for example, Michael Korey, *The Geometry of Power, the Power of Geometry: Mathematical Instruments and Princely Mechanical Devices from around 1600* (Munich, 2007).
5 Jim Bennett, 'A Role for Collections in the Research Agenda of the History of Science?', in *Research and Museums*, ed. Görel Cavalli-Björkman and Svante Lindqvist (Stockholm, 2008), pp. 193–210.
6 Silke Ackermann, '"Of Sufficient Interest . . ., but Not of Such Value . . .": 260 Years of Displaying Scientific Instruments in the British Museum', in *Scientific Instruments on Display*, ed. Silke Ackermann, Richard L. Kremer and Mara Miniati (Leiden, 2014), pp. 77–93.
7 Translated and quoted in Dominique Ferriot and Bruno Jacomy, 'The Musée des Arts et Métiers', in *Museums of Modern Science*, ed. Svante Lindqvist (Canton, MA, 2000), pp. 29–42 [p. 42].
8 Bernard V. Lightman, *Victorian Popularizers of Science: Designing Nature for New Audiences* (Chicago, IL, 2007).
9 Dominique Ferriot, 'The Role of the Object in Technical Museums: The Conservatoire National des Arts et Métiers', in *Museums and the Public Understanding of Science*, ed. John Durant (London, 1992), pp. 79–80.
10 Jeffrey A. Auerbach, *The Great Exhibition of 1851: A Nation on Display* (New Haven, CT, 1999); Jim Bennett, *Science at the Great Exhibition (London, 1851)* (Cambridge, 1983).
11 Robert Bud, 'Infected by the Bacillus of Science: The Explosion of South Kensington', in *Science for the Nation: Perspectives on the History of the Science Museum*, ed. Peter J. T. Morris (London, 2010), pp. 11–40.

12 Geoffrey N. Swinney, 'Towards an Historical Geography of a "National" Museum: The Industrial Museum of Scotland, the Edinburgh Museum of Science and Art and the Royal Scottish Museum, 1854–c. 1939', PhD thesis, University of Edinburgh, 2013.

13 Robert Bud, 'Responding to Stories: The 1876 Loan Collection of Scientific Apparatus and the Science Museum', *Science Museum Group Journal*, I (2014).

14 Brigitte Schroeder-Gudehus and Anne Rasmussen, *Les fastes du progrès: Le guide des expositions universelles, 1851–1992* (Paris, 1992); Canadelli, Beretta and Ronzon, eds, *Behind the Exhibit*.

15 Pamela M. Henson, '"Objects of Curious Research": The History of Science and Technology at the Smithsonian', *Isis*, XC (1999), pp. s249–69.

16 Deborah J. Warner, 'Joseph Henry and the Smithsonian's First Collection of Scientific Apparatus', *Scientific Instrument Society Bulletin*, CXLII (September 2019), pp. 26–32.

17 'Loan Collection of Scientific Apparatus', *Illustrated London News*, 16 September 1876, p. 270; Rebekah Higgitt, 'Instruments and Relics: The History and Use of the Royal Society's Object Collections, c. 1850–1950', *Journal of the History of Collections*, XXXI (2019), pp. 469–85.

18 Martin P. M. Weiss, *Showcasing Science: A History of Teylers Museum in the Nineteenth Century* (Amsterdam, 2019).

19 Bennett, 'A Role for Collections', p. 201.

20 Wolfgang M. Heckl, ed., *Technology in a Changing World: The Collections of the Deutsches Museum* (Munich, 2010); Eve M. Duffy, 'Representing Science and Technology: Politics and Display in the Deutsches Museum, 1903–1945', PhD thesis, University of North Carolina at Chapel Hill, 2002.

21 Lisa Kirch, *The Changing Face of Science and Technology in the Ehrensaal of the Deutsches Museum, 1903–1955* (Munich, 2017).

22 Swinney, 'Towards an Historical Geography of a "National" Museum'.

23 Museum of Science and Industry, *West Pavilion Guide* (Chicago, IL, 1938).

24 Klaus Staubermann and Geoffrey N. Swinney, 'Making Space for Models: (Re)presenting Engineering in Scotland's National Museum, 1854–Present', *International Journal for the History of Engineering and Technology*, LXXXVI (2016), pp. 19–41.

25 'New York City Museum of Science and Industry', *Nature*, CXXXVII (1936), p. 306.

26 Quotation from a 2008 interview with a visitor born in 1917 in Swinney, 'Towards an Historical Geography of a "National" Museum', p. 335.

27 Jim Bennett, 'European Science Museums and the Museum Boerhaave', in *75 jaar Museum Boerhaave* (Leiden, 2006), pp. 73–8; Marco Beretta, 'Andrea Corsini and the Creation of

the Museum of the History of Science in Florence, 1930–1961',
in *Scientific Instruments on Display*, ed. Silke Ackermann, Richard
L. Kremer and Mara Miniati (Leiden, 2014), pp. 1–36.

28 Sophie Forgan, 'Festivals of Science and the Two Cultures:
Science, Design and Display in the Festival of Britain, 1951', *British
Journal for the History of Science*, xxxi (1998), pp. 217–40.

29 Anne M. Zandstra and J. Wesley Null, 'How Did Museums Change
during the Cold War? Informal Science Education after Sputnik',
American Educational History Journal, xxxviii (2011), pp. 321–39;
Arthur P. Molella and Scott Gabriel Knowles, eds, *World's Fairs in
the Cold War: Science, Technology and the Culture of Progress*
(Pittsburgh, PA, 2019).

30 Arthur P. Molella, 'The Museum That Might Have Been: The
Smithsonian's National Museum of Engineering and Industry',
Technology and Culture, xxxii (1991), pp. 237–63.

31 'Johnson Dedicates Smithsonian Unit', *New York Times*,
23 January 1964, p. 28.

32 Arthur P. Molella, 'The Human Spirit in an Age of Machines', in
*World's Fairs in the Cold War: Science, Technology and the Culture
of Progress*, ed. Arthur P. Molella and Scott Gabriel Knowles
(Pittsburgh, PA, 2019), pp. 96–108.

33 United States National Museum, *Annual Report for the Year Ended
June 30, 1959* (Washington, DC, 1959), p. 3.

34 Bernard S. Finn, 'The Science Museum Today', *Technology and
Culture*, vi (1965), pp. 74–82.

35 John van Riemsdijk and Paul Sharp, *In the Science Museum*
(London, 1968); Connie Moon Sehat, 'Education and Utopia:
Technology Museums in Cold War Germany', PhD thesis, Rice
University, Houston, TX, 2006.

36 Jean-Baptiste Gouyon, '"Something Simple and Striking, if Not
Amusing": The Freedom 7 Special Exhibition at the Science
Museum, 1965', *Science Museum Group Journal*, i (2014).

37 John van Riemsdijk, *Science Museum: 50 Things to See*
(London, 1965).

38 Robert Hewison, *The Heritage Industry: Britain in a Climate of
Decline* (London, 1987).

39 Erin Beeston, 'Spaces of Industrial Heritage: A History of Uses,
Perceptions and the Re-Making of Liverpool Road Station,
Manchester', PhD thesis, University of Manchester, 2020.

40 K. C. Cole, *Something Incredibly Wonderful Happens: Frank
Oppenheimer and the World He Made Up* (Boston, MA, 2009).

41 Hilde Hein, *The Exploratorium: The Museum as Laboratory*
(Washington, DC, 1990), p. 85.

42 Frank Oppenheimer, 'Rationale for a Science Museum', *Curator:
The Museum Journal*, i (1968), pp. 206–9.

43 Karen A. Rader and Victoria E. M. Cain, *Life on Display:
Revolutionizing U.S. Museums of Science and Natural History
in the Twentieth Century* (Chicago, IL, 2014).

44 David Pantalony, conversation with the author, Edinburgh-Ottawa, 1 December 2020.

45 Victor J. Danilov, *Science and Technology Centres* (Cambridge, MA, 1982).

46 Alison Taubman, conversation with the author, Edinburgh, 11 December 2018. On Gregory helping Oppenheimer, see Richard Gregory, 'Turning Minds on to Science by Hands-On Exploration: The Nature and Potential of the Hands-On Medium', in Nuffield Foundation Interactive Science and Technology Project, *Sharing Science: Issues in the Development of Interactive Science and Technology Centres* (London, 1989), pp. 1–9.

47 Richard Gregory, *Hands-On Science: An Introduction to the Bristol Exploratory* (London, 1986).

48 Anthony Wilson, 'Launch Pad', in *Science Museum Review 1987*, ed. Andrew Nahum (London, 1987), pp. 22–5 [p. 22]; Tim Boon, 'Parallax Error? A Participant's Account of the Science Museum, c. 1980–c. 2000', in *Science for the Nation: Perspectives on the History of the Science Museum*, ed. Peter J. T. Morris (London, 2010), pp. 111–35.

49 National Museums of Scotland, *Annual Report April 1992–March 1993* (Edinburgh, 1993), p. 15.

50 Ian Simmons, 'A Conflict of Cultures: Hands-On Science Centres in UK Museums', in *Exploring Science in Museums*, ed. Susan Pearce (London, 1996), pp. 79–94.

51 Maurice Daumas, *Les instruments scientifiques aux XVIIe et XVIIIe siècles* (Paris, 1953); Alexandre Herlea, 'Maurice Daumas, 1910–1984', *Technology and Culture*, XXVI (1985), pp. 698–702.

52 Anthony V. Simcock, 'Alchemy and the World of Science: An Intellectual Biography of Frank Sherwood Taylor', *Ambix*, XXXIV (1987), pp. 121–39; Frank Greenaway, interview with Anna-K. Mayer, 5 March 1998, British Society for the History of Science Oral History Project, 'The History of Science in Britain, 1945–65', 31pp. TS transcript, BSHS 10/8/10, University of Leeds Special Collections.

53 Robert C. Post, *Who Owns America's Past? The Smithsonian and the Problem of History* (Baltimore, MD, 2013).

54 Jenni Calder and Alexander Fenton, eds, *National Museums of Scotland First Report, October 1985–March 1987* (Edinburgh, 1988).

55 Robert Fox, 'Museums of Science and Technology in Europe since 1980', in Frank Greenaway, *Chymica Acta: An Autobiographical Memoir*, ed. Robert G. W. Anderson, Peter J. T. Morris and Derek A. Robinson (Huddersfield, 2007), pp. 215–28.

56 Patricia E. Mooradian et al., *The Henry Ford* (Dearborn, MI, 2008).

57 Royal Society, *The Public Understanding of Science* (London, 1985), p. 9.

58 Nuffield Foundation Interactive Science and Technology Project, *Sharing Science: Issues in the Development of Interactive Science and Technology Centres* (London, 1989).

59 Danilov, *Science and Technology Centres*; Robert Fox, 'History and the Public Understanding of Science: Problems, Practices, and Perspectives', in *The Global and the Local: The History of Science and the Cultural Integration of Europe*, ed. Michal Kokowski (Cracow, 2006), pp. 174–7.

60 John R. Durant, Geoffrey A. Evans and Geoffrey P. Thomas, 'The Public Understanding of Science', *Nature*, cccxl (1989), pp. 11–14.

61 Steven Shapin, 'Why the Public Ought to Understand Science-in-the-Making', *Public Understanding of Science*, i (1992), pp. 27–30 [p. 30].

62 Julie Becker, '30 Years of Ecsite: Back to the Roots', *Spokes*, xlix (2019), pp. 1–12.

63 Rader and Cain, *Life on Display*; Committee on Prospering in the Global Economy of the 21st Century, *Rising above the Gathering Storm: Energizing and Employing America for a Brighter Economic Future* (Washington, dc, 2005).

64 Cited in Post, *Who Owns America's Past?*, p. 238.

65 Schäfer and Song, 'Interpreting the Collection'.

66 House of Lords, *Science and Technology: Third Report* (London, 2000), paragraph 3.40. See also Penny Fidler, 'Millennium Science Centres Historical Update', www.sciencecentres.org.uk, April 2020.

67 House of Lords, *Science and Technology*, paragraph 3.6.

68 Frank A.J.L. James, 'Some Significances of the Two Cultures Debate', *Interdisciplinary Science Reviews*, xli (2016), pp. 107–17 [p. 108].

69 Andrew Nahum, ed., *Science Museum Review 1987* (London, 1987), p. 5.

70 Sharon Macdonald, review of Svante Lindqvist, ed., *Museums of Modern Science* (Canton, ma, 2000), *British Journal for the History of Science*, xxxiv (2001), pp. 101–2.

2 Collecting Science

1 Alison Boyle, 'Of Mice and Myths: Challenges and Opportunities of Capturing Contemporary Science in Museums', *Science Museum Group Journal*, xiii (2020); Robert Bud, *The Uses of Life: A History of Biotechnology* (Cambridge, 1993).

2 Benjamin Filene, 'Things in Flux: Collecting in the Constructivist Museum', in *Active Collections*, ed. Elizabeth Wood, Rainey Tisdale and Trevor Jones (New York, 2018), pp. 130–40 [p. 130].

3 I surveyed four institutions' acquisitions in either 2017 or 2018: the Deutsches Museum via *Anhang zum Jahresbericht 2017* (Munich, 2018), provided by Helmuth Trischler; National Museums Scotland Science & Technology via our accessions register entries for 2018, as compiled by Julie Gibb; the Whipple Museum from an internal list provided by Josh Nall; and a list of Science Museum Group accessions made in 2018 provided by Jack Kirby. These data include some items that were already in

collections but had yet to be be formally accessioned. See also Tacye Phillipson, 'Collections Development in Hindsight: A Numerical Analysis of the Science and Technology Collections of National Museums Scotland since 1855', *Science Museum Group Journal*, XII (2019).

4 National Museum of American History, *Strategic Plan, 2013–2018* (Washington, DC, 2013), p. 6.

5 Richard L. Kremer, 'A Time to Keep, and a Time to Cast Away: Thoughts on Acquisitions for University Instrument Collections', *Rittenhouse*, XXII (2008), pp. 188–210.

6 'Elizabethan Combination Tide Computer', *Tesseract: Early Scientific Instruments*, CVI (2017–18), pp. 19–20.

7 Ingenium – Canada's Museums of Science and Innovation, *Collection Development Strategy* (Ottawa, 2018), pp. 1 and 2.

8 David Pantalony, 'Field Notes: Challenges and Approaches for Collecting Recent Material Heritage of Science and Technology', *Museologia e Patrimônio*, VIII (2015), pp. 80–103 [p. 90].

9 Sharon Macdonald and Jennie Morgan, 'What Not to Collect? Post-Connoisseurial Dystopia and the Profusion of Things', in *Curatopia: Museums and the Future of Curatorship*, ed. Philipp Schorch and Conal McCarthy (Manchester, 2019), pp. 29–43.

10 Paul Cornish, 'Extremes of Collecting at the Imperial War Museum, 1917–2009: Struggles with the Large and the Ephemeral', in *Extreme Collecting: Challenging Practices for 21st Century Museums*, ed. Graeme Were and J.C.H. King (New York, 2012), pp. 157–67.

11 Alison Boyle, 'Collecting and Interpreting Contemporary Science, Technology and Medicine at the Science Museum', in *Patrimoine contemporain des sciences et techniques*, ed. Catherine Ballé et al. (Paris, 2016), pp. 353–62.

12 Elsa Cox, *Age of Oil* (Edinburgh, 2017); Ellie Swinbank, 'Collecting and Displaying the Decommissioning of North Sea Oil and Gas at the National Museums Scotland', *Architectus*, LXI (2020), pp. 25–30.

13 Sarah Baines, 'From 2D to 3D: The Story of Graphene in Objects', *Science Museum Group Journal*, X (2018).

14 Pantalony, 'Field Notes: Challenges and Approaches', pp. 80–81.

15 Ibid., p. 90.

16 David Pantalony, 'Time-of-Flight Mass Spectrometer (TOF2)', unpublished acquisition proposal, Canada Science and Technology Museums Corporation, 19 November 2014, pp. 6 and 4; David Pantalony, 'Field Notes: Mass Spectrometry at the University of Manitoba', https://ingeniumcanada.org, 8 October 2019.

17 Elsa Cox, 'Energy Well Spent: Practical Approaches to Contemporary Collecting at the National Museum of Scotland', in *Patrimoine contemporain des sciences et techniques*, ed. Catherine Ballé et al. (Paris, 2016), pp. 321–30.

18 Kremer, 'A Time to Keep', p. 188.

19 Marta Lourenço, email to the author, 23 July 2017.
20 Tilly Blyth, 'Information Age? The Challenges of Displaying Information and Communication Technologies', *Science Museum Group Journal*, III (2015).
21 Anna Adamek, 'A Snapshot of Canadian Kitchens Collecting Contemporary Technologies as Historical Evidence for Future Research', in *Challenging Collections: Approaches to the Heritage of Recent Science and Technology*, ed. Alison Boyle and Johannes-Geert Hagmann (Washington, DC, 2017), pp. 134–49.
22 National Museum of American History, 'National Museum of American History Implements Collecting Strategy in Response to COVID-19 Pandemic', https://americanhistory.si.edu, 8 April 2020.
23 Robert G. W. Anderson, *The Playfair Collection* (Edinburgh, 1978).
24 Susan M. Pearce, *On Collecting: An Investigation into Collecting in the European Tradition* (London, 1995), p. 407.
25 'Knick-knackatory' was a derisive term for a cabinet of curiosities, a collection of knick-knacks. This was in a *Universal Magazine* satire about Hans Sloane, whose collections form the nucleus of the British Museum, likely to have been penned by antiquarian and librarian Thomas Hearne (1678–1735). Cited by W. D. Ian Rolfe in J. M. Chalmers-Hunt, ed., *Natural History Auctions, 1700–1972: A Register of Sales in the British Isles* (London, 1976), p. 36.
26 Alan Q. Morton and Jane A. Wess, *Public and Private Science: The King George III Collection* (Oxford, 1993); Alexandra Rose and Jane Desborough, *Science City: Craft, Commerce and Curiosity in London, 1550–1800* (London, 2020).
27 Sebastian Chan and Aaron Cope, 'Planetary: Collecting and Preserving Code as a Living Object', www.cooperhewitt.org, 26 August 2013; Petrina Foti, *Collecting and Exhibiting Computer-Based Technology* (Abingdon, 2019).
28 Science Museum, 'Superbugs: The Fight for Our Lives', www.sciencemuseum.org.uk, 9 November 2017.
29 Clifford Lynch, 'Stewardship in the "Age of Algorithms"', https://firstmonday.org, 4 December 2017.
30 Yuk Hui, *On the Existence of Digital Objects* (Minneapolis, MN, 2016), p. 5.
31 Henry Lowood, 'Defining the Software Collection', in *Challenging Collections: Approaches to the Heritage of Recent Science and Technology*, ed. Alison Boyle and Johannes-Geert Hagmann (Washington, DC, 2017), pp. 68–86.
32 Ross Parry, 'The End of the Beginning: Normativity in the Postdigital Museum', *Museum Worlds*, I (2013), pp. 24–39.
33 Quoted in John E. Simmons, *Things Great and Small: Collection Management Policies* (Washington, DC, 2006), p. 51.
34 For a recent discussion, see, for example, Jennie Morgan and Sharon Macdonald, 'De-Growing Museum Collections for New Heritage Futures', *International Journal of Heritage Studies*, XXVI (2020), pp. 56–70.

35   Robert Bud, 'Collecting for the Science Museum: Constructing the Collections, the Culture and the Institution', in *Science for the Nation: Perspectives on the History of the Science Museum*, ed. Peter J. T. Morris (London, 2010), pp. 266–88.

36   Phillipson, 'Collections Development in Hindsight'.

37   National Maritime Museum, 'National Maritime Museum Collections Reform Project', www.rmg.co.uk, 30 November 2004.

38   Scottish Transport and Industrial Collections Knowledge Network, 'Old Tools, New Uses', http://stickssn.org, October 2011.

39   National Air and Space Museum, '2018 – NASM Objects Available for Transfer', https://airandspace.si.edu, 21 December 2018.

40   Ingenium – Canada's Museums of Science and Innovation, *Annual Report, 2017–18* (Ottawa, 2018).

41   Anonymous curator, quoted in Harald Fredheim, Sharon Macdonald and Jennie Morgan, *Profusion in Museums: A Report on Contemporary Collecting and Disposal* (York, 2018), p. 21.

42   Science Museum Group, *Collection Development*, p. 3.

43   Pantalony, 'Time-of-Flight Mass Spectrometer (TOF2)', p. 4.

44   David Pantalony, email to the author, 3 December 2020.

3 Treasures of the Storeroom

1   James D. Inglis, 'Typewriters and Commerce in Scotland, 1875–1930', PhD thesis, University of St Andrews, 2022; on Smithies, see Malcolm Robert Petrie, 'Public Politics and Traditions of Popular Protest: Demonstrations of the Unemployed in Dundee and Edinburgh, c. 1921–1939', *Contemporary British History*, XXVII (2013), pp. 490–513.

2   Hilary Geoghegan and Alison Hess, 'Object-Love at the Science Museum: Cultural Geographies of Museum Storerooms', *Cultural Geographies*, XXII (2015), pp. 445–65 [p. 451]. Other scholars shifting academic attention from exhibition and collecting to museum storage include Mirjam Brusius and Kavita Singh, eds, *Museum Storage and Meaning: Tales from the Crypt* (Abingdon, 2018); Stefan Oláh and Martina Griesser-Stermscheg, eds, *Museumsdepots: Inside the Museum Storage* (Salzburg, 2014); an autobiographical account can be found in Richard Fortey, *Dry Store Room No. 1: The Secret Life of the Natural History Museum* (London, 2008).

3   For broad statistics see Suzanne Keene, ed., *Collections for People: Museums' Stored Collections as a Public Resource* (London, 2008); Heritage Preservation and the Institute of Museum and Library Services, *A Public Trust at Risk: The Heritage Health Index Report on the State of America's Collections* (Washington, DC, 2005).

4   GWP Architecture, *Science Museum Group – Building One Collections Storage Facility: Pre-Application Enquiry* (London, 2017).

5　Sonia Mendes, 'Ingenium's Collections Conservation Centre: Bringing Canada's Past into the Future', *Muse Magazine* (March–April 2019), pp. 24–31.

6　Marie Grima, 'The Tod Head Lighthouse Lantern', *Architectus*, LXI (2020), pp. 9–16.

7　Association of British Transport and Engineering Museums, *Guidelines for the Care of Larger and Working Historic Objects* (London, 2018).

8　Andrew Howe and Jacek Wiklo, 'Riverside Museum: Building a New State of the Art Transport and Technology Museum in Glasgow, Scotland', http://bigstuff.omeka.net, 2013.

9　Sharon Macdonald, *Behind the Scenes at the Science Museum* (London, 2002).

10　Sharon Macdonald, 'Museum Storerooms', https://heritage-futures. org, accessed 19 December 2021.

11　See, for example, Janine Fox, 'One Year On: A Move Project Team Update', https://blogs.mhs.ox.ac.uk, 31 August 2017.

12　HM Treasury, *Spending Review and Autumn Statement 2015* (London, 2015).

13　Alison Morrison-Low, book review of Gerard L'Estrange Turner, *The Practice of Science in the Nineteenth Century*, *Technology and Culture*, XXXIX (1998), pp. 563–4 [p. 563].

14　Louis Volkmer, conversation with the author, Edinburgh, 6 March 2018; he had recently returned from the international seminar 'Material Culture in the History of Physics' at the Deutsches Museum, 26 February–2 March 2018.

15　Nicholas Thomas, *The Return of Curiosity: What Museums Are Good For in the 21st Century* (London, 2016).

16　James Inglis, email to the author, 3 July 2018. See Inglis, 'Typewriters and Commerce in Scotland'.

17　Steven Lubar, *Inside the Lost Museum: Curating, Past and Present* (Cambridge, MA, 2017).

18　David Pantalony, 'Collectors, Displays and Replicas in Context: What We Can Learn from Provenance Research in Science Museums', in *The Romance of Science*, ed. Jed Buchwald and Larry Stewart (Cham, 2017), pp. 255–75.

19　Thomas, *The Return of Curiosity*.

20　Katharine Anderson et al., 'Reading Instruments: Objects, Texts and Museums', *Science and Education*, XXII (2013), pp. 1167–89 [p. 1173]. See also David Pantalony, 'What Remains: The Enduring Value of Museum Collections in the Digital Age', *HOST: Journal of History of Science and Technology*, XIV (2020), pp. 160–82.

21　For the period 2007–17, Louis Volkmer surveyed the history of science journal *Isis* and James Inglis looked at *Technology and Culture*. Only eleven of the 343 articles showed evidence of the authors directly experiencing museum objects. See Samuel J.M.M. Alberti, Alison Boyle, James Inglis and Louis Volkmer, 'The Immaterial Turn? How Historians of Science and Technology

Use Material Culture', in *Understanding Use*, ed. Tim Boon et al. (Washington, DC, forthcoming).

22 Elizabeth Haines and Anna Woodham, 'Mobilising the Energy in Store', *Science Museum Group Journal*, XII (2019); Anna Woodham, Alison Hess and Rhianedd Smith, eds, *Exploring Emotion, Care, and Enthusiasm in 'Unloved' Museum Collections* (Leeds, 2020).

23 Simon Stephens, 'The Art of Science', *Museums Journal* (February 2012), pp. 34–7 [p. 37].

24 The former an unnamed museum, possibly apocryphal; the latter the Thinktank, Birmingham Science Museum, courtesy of Jack Kirby.

25 Martha Fleming, 'People, Places and Things: New Models for Collections-Based Research', www.vam.ac.uk, 20 March 2017.

26 See, for example, Richard Dunn and Rebekah Higgitt, *Finding Longitude: How Clocks and Stars Helped Solve the Longitude Problem* (Glasgow, 2014).

27 Henrik Treimo, 'Sketches to a Methodology for Museum Research', in *Exhibitions as Research: Experimental Methods in Museums*, ed. Peter Bjerregaard (Abingdon, 2020), pp. 19–39.

28 Science Museum Group, 'Research Strategy', www.sciencemuseumgroup.org.uk, 2018.

29 Nicky Reeves, 'Visible Storage, Visible Labour?', in *Museum Storage and Meaning: Tales from the Crypt*, ed. Mirjam Brusius and Kavita Singh (Abingdon, 2018), pp. 55–63 [p. 58]; Thomas Thiemeyer, 'The Storeroom as Promise: The Discovery of the Ethnological Museum Depot as an Exhibition Method in the 1970s', *Museum Anthropology*, XL (2017), pp. 143–57.

30 Trevor Jones, 'A (Practical) Inspiration: Do You Know What It Costs You to Collect?', in *Active Collections*, ed. Elizabeth Wood, Rainey Tisdale and Trevor Jones (New York, 2018), pp. 141–4; Nick Merriman, 'Museum Collections and Sustainability', *Cultural Trends*, XVII (2008), pp. 3–21. I thank Klaus Staubermann for the phrase 'fragments of eternity'.

31 Ólöf Gerður Sigfúsdóttir, 'Blind Spots: Museology on Museum Research', *Museum Management and Curatorship*, XXXV (2020), pp. 196–209.

4 Engaging Objects

1 Nanoscale Informal Science Education Network, 'Exploring Materials – Ferrofluid', www.nisenet.org, 2013.

2 National Informal STEM Education Network, *Report to Partners 2005–2016* (Boston, MA, 2017).

3 Ecsite, *Together: Annual Report* (Brussels, 2018).

4 Science Museum Group, Written Evidence to Science and Technology Committee, http://data.parliament.uk, April 2016, paragraph 3.1.

5 Rae Ostman, *Nano Exhibition: Creating a Small-Footprint Exhibition with a Big Impact* (Saint Paul, MN, 2015), p. 7; Rae Ostman and Catherine McCarthy, 'Nano: Creating an Exhibition That Is Inclusive of Multiple and Diverse Audiences', *Exhibitionist*, XXXIV (Fall 2015), pp. 34–9.

6 Sandra Murriello and Marcelo Knobel, 'NanoAdventure: An Interactive Exhibition in Brazil', in *Science Exhibitions: Curation and Design*, ed. Anastasia Filippoupoliti (Edinburgh, 2010), pp. 394–414; Paul A. Youngman and Ljiljana Fruk, 'A Nanochemist and a Nanohumanist Take a Walk through the German Museum: An Analysis of the Popularization of Nanoscience and Technology in Germany', *Journal of Conservation and Museum Studies*, XII (2014), pp. 1–8.

7 Author's fieldwork, Science Museum, London, 24 January 2020; Alexandra Rose and Jane Desborough, *Science City: Craft, Commerce and Curiosity in London, 1550–1800* (London, 2020), p. 20.

8 Emily Cronin, 'The Future of Travelling Exhibitions', *Spokes*, LII (May 2019).

9 Melanie Jahreis, Sara Marquart and Nina Möllers, eds, *Kosmos Kaffee* (Munich, 2019).

10 Henrik Treimo, 'Mind Gap', *Interdisciplinary Science Reviews*, XXXVIII (2013), pp. 259–74 [pp. 259, 267–8].

11 As helpfully articulated by the Exploratorium: Kathleen McLean and Catherine McEver, eds, *Are We There Yet? Conversations about Best Practices in Science Exhibition Development* (San Francisco, CA, 2004).

12 Helen Graham, 'The "Co" in Co-Production: Museums, Community Participation and Science and Technology Studies', *Science Museum Group Journal*, V (2016).

13 Alison Boyle, 'SIS Members Are Instrumental in New Science Museum Gallery', *Bulletin of the Scientific Instrument Society*, CXLII (2019), pp. 33–4.

14 Hope-Stone Research and National Museums Scotland, 'Parasites Exhibition: Audience Research Report', May 2020, p. 73, National Museums Scotland unpublished evaluation.

15 Treimo, 'Mind Gap', pp. 268–9.

16 Elizabeth Jones, from *Genome: Unlocking Life's Code*, National Museum of Natural History, 2013, cited in American Alliance of Museums, *Excellence in Label Writing, 2014*, www.aam-us.org, 2014.

17 Joshua P. Gutwill and Toni Dancstep, 'Boosting Metacognition in Science Museums: Simple Exhibit Label Designs to Enhance Learning', *Visitor Studies*, XX (2017), pp. 72–88.

18 Haidy Geismar, *Museum Object Lessons for the Digital Age* (London, 2018); Cooper Hewitt, Smithsonian Design Museum, 'Using the Pen', www.cooperhewitt.org, accessed 22 December 2020.

19 Author's fieldwork, Harvard University, Cambridge, MA,
    3 November 2017; Alvin Powell, 'Galileo to Cyclotron: History on
    Display', *Harvard Gazette*, 15 December 2005.

20 Rebekah Higgitt, 'Challenging Tropes: Genius, Heroic Invention,
    and the Longitude Problem in the Museum', *Isis*, CVIII (2017),
    pp. 371–80.

21 Ian Blatchford, 'If Not Now, When?', www.sciencemuseumgroup.
    org.uk, 12 June 2020.

22 Elaine Heumann Gurian, 'Offering Safer Public Spaces', *Journal of
    Museum Education*, XXI (1995), pp. 14–16 [p. 14]. On controversial
    exhibitions, see Erminia Pedretti and Ana Maria Navas Iannini,
    *Controversy in Science Museums: Re-Imagining Exhibition Spaces
    and Practice* (Abingdon, 2020).

23 Author's fieldwork, Whipple Museum of the History of Science,
    University of Cambridge, 8 May 2019.

24 Rosanna Evans, email to the author, 17 May 2019.

25 Kevin Crowley and Melanie Jacobs, 'Building Islands of Expertise
    in Everyday Family Activity', in *Learning Conversations in
    Museums*, ed. Gaea Leinhardt, Kevin Crowley and Karen Knutson
    (Mahwah, NJ, 2002), pp. 333–56 [p. 333]; Stephan Schwan,
    Alejandro Grajal and Doris Lewalter, 'Understanding and
    Engagement in Places of Science Experience: Science Museums,
    Science Centers, Zoos, and Aquariums', *Educational Psychologist*,
    XLIX (2014), pp. 70–85.

26 Rosanna Evans, email to the author, 30 May 2019.

27 Science Museum Group, *Annual Report and Accounts, 2019–20*
    (London, 2020).

28 Linda Weintraub, 'SUPERFLEX – Join a Cockroach Tour of a Science
    Museum', http://lindaweintraub.com, 20 November 2015.

29 Christine Reich et al., 'NISE NET: Team-Based Inquiry', in *The
    Reflective Museum Practitioner: Expanding Practice in Science
    Museums*, ed. Laura W. Martin, Lynn Uyen Tran and Doris B. Ash
    (Abingdon, 2019), pp. 53–63.

30 Jack Stilgoe, *Nanodialogues: Experiments in Public Engagement
    with Science* (London, 2007), p. 13; Barbara N. Flagg and Valerie
    Knight-Williams, *Summative Evaluation of NISE Network's Public
    Forum: Nanotechnology in Health Care* (Bellport, NY, 2008).

31 Wiktor Gajewski, 'After-Hours Events', *Spokes*, LXI (March 2020).

32 Ellen McCallie et al., 'Learning to Generate Dialogue: Theory,
    Practice, and Evaluation', *Museums and Social Issues*, II (2007),
    pp. 165–84 [p. 165].

33 Jessica Brown, 'Man Accidentally Starts Twitter War between
    Natural History and Science Museums', www.indy100.com,
    16 September 2017.

34 John Stack, 'How Museums Have Been Transformed by the Digital
    Revolution', www.jisc.ac.uk, 26 February 2019.

35 Mitchell Whitelaw, 'Generous Interfaces for Digital Cultural
    Collections', *Digital Humanities Quarterly*, IX (2015), paragraph 46.

36 Science Museum Group, 'Search Our Collection', http://collection. sciencemuseum.org.uk, accessed 31 December 2021.

37 Kira Zumkley, 'Taking Collection Digitisation to the Next Level', www.sciencemuseumgroup.org.uk, 2 October 2018.

38 Stack, 'How Museums Have Been Transformed'.

39 Ian Sample, 'The Royal Tweet: Queen Sends First Twitter Message', *The Guardian*, 24 October 2014; Valentine Low, 'Queen Posts First Instagram Photo at Science Museum', *The Times*, 7 March 2019.

40 As of 30 June 2019 I charted the five most recent blogs, Facebook posts, Instagram posts, videos and stories, YouTube videos and any tweets over the previous five days of the Canada Museum of Science and Technology (rather than Ingenium, its parent body); Deutsches Museum; Exploratorium (a science centre for comparison); National Museums Scotland (a multidisciplinary museum for comparison, but charting only content relating to the Science and Technology collection); and the Science Museum (specifically the London site rather than the Science Museum Group). In total I reviewed 165 posts in relation to promotion, activities, exhibitions, object images, staff activity, museum practice and 'on this day' historical events.

41 Paige Brown Jarreau, Nicole Smith Dahmen and Ember Jones, 'Instagram and the Science Museum: A Missed Opportunity for Public Engagement', *Journal of Science Communication*, XVIII (2019), A06, pp. 1–22.

42 Boris Jardine, Joshua Nall and James Hyslop, 'More than Mensing? Revisiting the Question of Fake Scientific Instruments', *Bulletin of the Scientific Instrument Society*, CXXXII (2017), pp. 22–9.

43 Katie Birkwood, 'Make Your Own Anatomical Manikin: Human Anatomy Model Inspired by Andreas Vesalius', https://history. rcplondon.ac.uk, 17 April 2020.

44 Paul R. Brewer and Barbara L. Ley, '"Where My Ladies At?" Online Videos, Gender, and Science Attitudes among University Students', *International Journal of Gender, Science and Technology*, IX (2018), pp. 278–97.

45 Russell Dornan, 'Should Museums Have a Personality?', https:// medium.com, 9 March 2017.

46 Vickie Curtis, *Parasites: Battle for Survival* ethnographic visitor observation, 29 February 2020, National Museums Scotland; Hope-Stone Research and National Museums Scotland, '*Parasites* Exhibition: Audience Research Report', May 2020, p. 21, National Museums Scotland unpublished evaluation.

47 Jenny Kidd, 'Digital Media Ethics and Museum Communication', in *The Routledge Handbook of Museums, Media and Communication*, ed. Kirsten Drotner et al. (London, 2019), pp. 193–204 [p. 195].

48 There is a large literature on learning theory as applied to museums: see, for example, John H. Falk and Lynn D. Dierking, *Learning from Museums*, 2nd edn (Lanham, MD, 2018).

49 Neta Shaby, Orit Ben-Zvi Assaraf and Tali Tal, 'Engagement in a Science Museum: The Role of Social Interactions', *Visitor Studies*, XXII (2019), pp. 1–20.

50 Pedretti and Navas Iannini, *Controversy in Science Museums*, p. 168; Sarah R. Davies, 'Knowing and Loving: Public Engagement beyond Discourse', *Science and Technology Studies*, XXVII (2014), pp. 90–110.

51 David Holdsworth, 'History, Nostalgia and Software', in *Making the History of Computing Relevant*, ed. Arthur Tatnall, Tilly Blyth and Roger Johnson (Heidelberg, 2013), pp. 266–73.

52 Author's fieldwork, Science Museum, London, 24 January 2020.

53 WhatsApp message to the author, 29 May 2020, original emphasis.

54 Falk and Dierking, *Learning from Museums*.

55 Paul DeHart Hurd, 'Science Literacy: Its Meaning for American Schools', *Educational Leadership*, XVI (1958), pp. 13–16 and 52; for a round-up of its development and deployment in science museums, see Pedretti and Navas Iannini, *Controversy in Science Museums*.

56 Louise Archer et al., *Science Capital Made Clear* (London, 2016), p. 2. Louise Archer led the programme known as ASPIRES who developed these concepts. For more detail, see Louise Archer et al., '"Science Capital": A Conceptual, Methodological, and Empirical Argument for Extending Bourdieusian Notions of Capital beyond the Arts', *Journal of Research in Science Teaching*, LII (2015), pp. 922–48.

57 Elizabeth Kunz Kollmann et al., NISE *Net Research on How Visitors Find and Discuss Relevance in the Nano Exhibition* (Boston, MA, 2015), p. 5.

58 Ipsos MORI, *Public Attitudes to Science, 2014* (London, 2014).

59 Hope-Stone Research and National Museums Scotland, 'Get Energised 2018–19 Evaluation Findings', May 2019, p. 34; '*Parasites* Exhibition: Audience Research Report', May 2020, p. 74, National Museums Scotland unpublished evaluation.

60 Constanze Hampp and Stephan Schwan, 'The Role of Authentic Objects in Museums of the History of Science and Technology: Findings from a Visitor Study', *International Journal of Science Education*, Part B, V (2015), pp. 161–81.

61 Science Museum Group, Written Evidence to Science and Technology Committee, http://data.parliament.uk, April 2016, paragraph 4.3.

62 Robin Boast, 'Neocolonial Collaboration: Museum as Contact Zone Revisited', *Museum Anthropology*, XXXIV (2011), pp. 56–70.

63 David Chittenden, 'Roles, Opportunities, and Challenges: Science Museums Engaging the Public in Emerging Science and Technology', *Journal of Nanoparticle Research*, XIII (2011), pp. 1549–56 [p. 1550].

64 Sharon Macdonald, 'Exhibition Experiments: Publics, Politics and Scientific Controversy', in *Science Exhibitions: Curation and Design*, ed. Anastasia Filippoupoliti (Edinburgh, 2010), pp. 138–51 [p. 141].

5 Campaigning with Collections

1 Nina Möllers, Christian Schwägerl and Helmuth Trischler, eds, *Welcome to the Anthropocene: The Earth in Our Hands* (Munich, 2015); Deutsches Museum, 'Die Traktoren kommen', www.instagram.com accessed 15 April 2022.

2 Deutsches Museum, 'Leitbild', www.deutsches-museum.de, accessed 19 December 2021.

3 See, for example, Fiona Cameron, 'Young People, Climate Mobilisation and Science Centre Alliances', *Spokes*, CXVIII (2020); and Ipsos MORI, *Veracity Index 2020: Trust in Professions Survey* (London, 2020), in which only medics, engineers, teachers, judges, scientists and professors are trusted more than curators of the thirty categories surveyed.

4 Elaine Heumann Gurian, 'Offering Safer Public Spaces', *Journal of Museum Education*, XXII (1995), pp. 14–16 [p. 14].

5 Anonymized respondent, cited in Marianne Achiam and Jan Sølberg, 'Nine Meta-Functions for Science Museums and Science Centres', *Museum Management and Curatorship*, XXXII (2017), pp. 123–43 [p. 136].

6 Erminia Pedretti and Ana Maria Navas Iannini, *Controversy in Science Museums: Re-Imagining Exhibition Spaces and Practice* (Abingdon, 2020).

7 This articulation of activism is drawn from Sandra L. Rodegher and Stacey Vicario Freeman, 'Advocacy and Action', in *Museum Activism*, ed. Robert R. Janes and Richard Sandell (Abingdon, 2019), pp. 337–47.

8 Data from the U.S. National Awareness, Attitudes, and Usage Study, cited in Colleen Dilenschneider, 'People Trust Museums More than Newspapers', www.colleendilen.com, 26 April 2017; see also Fiona R. Cameron, 'Climate Change, Agencies, and the Museum for a Complex World', *Museum Management and Curatorship*, XXVII (2012), pp. 317–39.

9 Karen Knutson, 'Rethinking Museum/Community Partnerships', in *The Routledge Handbook of Museums, Media and Communication*, ed. Kirsten Drotner et al. (London, 2019), pp. 101–14.

10 Jennifer Newell, 'Talking around Objects: Stories for Living with Climate Change', in *Curating the Future: Museums, Communities and Climate Change*, ed. Jennifer Newell, Libby Robin and Kirsten Wehner (Abingdon, 2017), pp. 34–49.

11 Ingenium – Canada's Museums of Science and Innovation, 'From Earth to Us', https://ingeniumcanada.org, accessed 13 November 2020; Katherine Anderson and Jan Hadlaw,

'The Canada Science and Technology Museum', *Technology and Culture*, LIX (2018), pp. 781–6.

12 Möllers, *Welcome to the Anthropocene*, pp. 130, 131. See also Lotte Isager, Line Vestergaard Knudsen and Ida Theilade, 'A New Keyword in the Museum: Exhibiting the Anthropocene', *Museum and Society*, XIX (2021), pp. 88–107.

13 Nina Möllers, Luke Keogh and Helmuth Trischler, 'A New Machine in the Garden? Staging Technospheres in the Anthropocene', in *Gardens and Human Agency in the Anthropocene*, ed. Maria Paula Diogo et al. (Abingdon, 2019), pp. 161–79 [p. 166].

14 Nina Möllers, 'Cur(at)ing the Planet: How to Exhibit the Anthropocene and Why', RCC *[Rachel Carson Centre] Perspectives*, III (2013), pp. 57–66 [p. 63].

15 Möllers, *Welcome to the Anthropocene*, p. 123; cf. Cameron, 'Climate Change'.

16 Möllers, 'Cur(at)ing the Planet', p. 60.

17 Finn Arne Jørgensen and Dolly Jørgensen, 'The Anthropocene as a History of Technology: *Welcome to the Anthropocene: The Earth in Our Hands*, Deutsches Museum, Munich', *Technology and Culture*, LVII (2016), pp. 231–7 [p. 233].

18 Anthony Leiserowitz and Nicholas Smith, *Knowledge of Climate Change among Visitors to Science and Technology Museums* (New Haven, CT, 2011).

19 Robert R. Janes, 'The End of Neutrality: A Modest Manifesto', *Informal Learning Review*, CXXXV (2015), pp. 3–8.

20 Mel Evans, *Artwash: Big Oil and the Arts* (London, 2015).

21 Steve Bird, 'Student Climate Change Activist Plan to Target Science Museum over Oil Sponsorship', *Daily Telegraph*, 5 October 2019.

22 Ian Blatchford, 'Wonderlab: The Statoil Gallery at the Science Museum', *Spokes*, XXIV (2016); Ian Blatchford, 'Setting Out Our Approach to the Century's Defining Challenge', www.sciencemuseumgroup.org.uk, 18 November 2020.

23 Science Museum Group, 'Science Museum Group Announces Major Public Programme on Climate Change', www.sciencemuseum.org.uk, 4 February 2020.

24 Beginning with Roger Highfield, 'Coronavirus: What We Know (and Don't Know) about the Virus', www.sciencemuseumgroup. org.uk, 23 March 2020.

25 Natural History Museum, 'Museum Botanist Dr Sandy Knapp has Studied Plants for 40 Years', www.instagram.com, 7 April 2020. The post remains, but the exchange has since been removed.

26 Quoted in Michael Creek et al., eds, *Tackling Misinformation: A Collection of Research and Resources for Science Engagement Professionals Addressing the Spread of Inaccurate Information about Science and Scientists*, www.ecsite.eu, 2020. p. 20.

27 António Gomes da Costa, 'From Ear Candling to Trump: Science Communication in the Post-Truth World', *Spokes*, XXVII (2017).

28 Richard Sandell, *Museums, Moralities and Human Rights* (Abingdon, 2017).

29 Katherine Ott, 'To Junius Wilson, Bikes Meant Freedom', https://americanhistory.si.edu, 23 July 2015; Susan Burch and Hannah Joyner, *Unspeakable: The Story of Junius Wilson* (Chapel Hill, NC, 2007).

30 Pallavi Podapai, 'Exhibit Lab at the Science History Institute', http://allofusdha.org, 18 November 2019.

31 Quoted in Sophie Goggins, Tacye Phillipson and Samuel J.M.M. Alberti, 'Prosthetic Limbs on Display: From Maker to User', *Science Museum Group Journal*, VIII (2017).

32 Stephen Mullen, 'The Rise of James Watt', in *James Watt (1736–1819): Culture, Innovation and Enlightenment*, ed. Caroline Archer-Parré and Malcolm Dick (Liverpool, 2020), pp. 39–60; Eric Williams, *Capitalism and Slavery* (Chapel Hill, NC, 1944).

33 Chris Flink, 'Black Lives Matter', www.exploratorium.edu, 8 June 2020; Christina Tessier, 'Ingenium's Response to Recent Racial Violence', https://ingeniumcanada.org, accessed 1 November 2020.

34 Ian Blatchford, 'If Not Now, When?', www.sciencemuseumgroup.org.uk, 12 June 2020; Nadine White, 'Science Museum Criticised by Staff over Lack of Response to Black Lives Matter Movement', *Huffington Post*, 12 June 2020.

35 Tilly Blyth, 'Rethinking Collections Research', www.sciencemuseumgroup.org.uk, 10 July 2020.

36 Ibid.

37 Bernadette T. Lynch and Samuel J.M.M. Alberti, 'Legacies of Prejudice: Racism, Co-Production and Radical Trust in the Museum', *Museum Management and Curatorship*, XXV (2010), pp. 13–35; see also Richard Sandell, *Museums, Prejudice and the Reframing of Difference* (Abingdon, 2007).

38 Samuel J.M.M. Alberti, 'Museum Nature', in *Worlds of Natural History*, ed. Helen Ann Curry et al. (Cambridge, 2018), pp. 348–62.

39 Flink, 'Black Lives Matter'.

40 Hooley McLaughlin, 'Youth and the Challenge to Define the Museum of the Future', in *Science Museums in Transition: Unheard Voices*, ed. Hooley McLaughlin and Judy Diamond (Abingdon, 2020), pp. 80–89 [p. 81].

41 Anonymous visitor, interviewed and quoted by Pedretti and Navas Iannini, *Controversy in Science Museums*, p. 169.

42 Dominic Berry, 'Review: FOLK Exhibition, Oslo', *Viewpoint*, CXVIII (2019), p. 14.

43 Kaja Hannedatter Sontum, 'The Co-Production of Difference? Exploring Urban Youths' Negotiations of Identity in Meeting with Difficult Heritage of Human Classification', *Museums and Social Issues*, XIII (2018), pp. 43–57 [p. 43].

44 Flink, 'Black Lives Matter'.

45 Science and Industry Museum, '#bornthisday Marie Van Brittan Brown', https://twitter.com/sim_manchester, 30 October 2020;

Camille Leadbetter, 'First Impressions of the Portrait of Sir John Chardin', https://blogs.mhs.ox.ac.uk, 15 December 2020.

46 Margot Lee Shetterly, *Hidden Figures: The Story of the African-American Women Who Helped Win the Space Race* (London, 2016); Ellen Stofan, 'Remembering Katherine Johnson: NASA Mathematician Calculated Mission Flight Paths and Continues to Inspire', https://airandspace.si.edu, 25 February 2020.

47 Flink, 'Black Lives Matter'.

48 Anonymous visitor, interviewed and quoted by Pedretti and Navas Iannini, *Controversy in Science Museums*, pp. 169 and 176.

6 Lively Collections

1 Tilly Blyth, 'Rethinking Collections Research', www.sciencemuseumgroup.org.uk, 10 July 2020.

2 Anonymous curator, interviewed and quoted in Erminia Pedretti and Ana Maria Navas Iannini, *Controversy in Science Museums: Re-Imagining Exhibition Spaces and Practice* (Abingdon, 2020), pp. 101–2.

3 Trevor Jones and Rainey Tisdale, 'A Manifesto for Active History Museum Collections', in *Active Collections*, ed. Elizabeth Wood, Rainey Tisdale and Trevor Jones (New York, 2018), pp. 7–10.

4 Bill Watson and Shari Rosenstein Werb, 'One Hundred Strong: A Colloquium on Transforming Natural History Museums in the Twenty-First Century', *Curator: The Museum Journal*, CVI (2013), pp. 255–65; Karen Knutson, 'Rethinking Museum/Community Partnerships', in *The Routledge Handbook of Museums, Media and Communication*, ed. Kirsten Drotner et al. (London, 2019), pp. 101–14.

5 Nina Simon, *The Art of Relevance* (Santa Cruz, CA, 2016). Marianne Achiam and Jan Sølberg, 'Nine Meta-Functions for Science Museums and Science Centres', *Museum Management and Curatorship*, XXXII (2017), pp. 123–43.

SELECT BIBLIOGRAPHY

Achiam, Marianne, and Jan Solberg, 'Nine Meta-Functions for Science Museums and Science Centres', *Museum Management and Curatorship*, XXXII (2017), pp. 123–43

Alberti, Samuel J.M.M., and Elizabeth Hallam, eds, *Medical Museums: Past, Present, Future* (London, 2013)

Anderson, Katharine, Mélanie Frappier, Elizabeth Neswald and Henry Trim, 'Reading Instruments: Objects, Texts and Museums', *Science and Education*, XXII (2013), pp. 1167–89

Ballé, Catherine, et al., eds, *Patrimoine contemporain des sciences et techniques* (Paris, 2016)

Bjerregaard, Peter, ed., *Exhibitions as Research: Experimental Methods in Museums* (Abingdon, 2020)

Boyle, Alison, and Johannes-Geert Hagmann, eds, *Challenging Collections: Approaches to the Heritage of Recent Science and Technology* (Washington, DC, 2017)

Brusius, Mirjam, and Kavita Singh, eds, *Museum Storage and Meaning: Tales from the Crypt* (Abingdon, 2018)

Canadelli, Elena, Marco Beretta and Laura Ronzon, eds, *Behind the Exhibit: Displaying Science and Technology at World's Fairs and Museums in the Twentieth Century* (Washington, DC, 2019)

Cavalli-Björkman, Görel, and Svante Lindqvist, eds, *Research and Museums* (Stockholm, 2008)

Dawson, Emily, *Equity, Exclusion and Everyday Science Learning: The Experiences of Minoritised Groups* (London, 2019)

Filippoupoliti, Anastasia, ed., *Science Exhibitions*, 2 vols (Edinburgh, 2010)

Fortey, Richard, *Dry Store Room No. 1: The Secret Life of the Natural History Museum* (London, 2008)

Geoghegan, Hilary, and Alison Hess, 'Object-Love at the Science Museum: Cultural Geographies of Museum Storerooms', *Cultural Geographies*, XXII (2015), pp. 445–65

Gutwill, Joshua P., and Sue Allen, *Group Inquiry at Science Museum Exhibits: Getting Visitors to Ask Juicy Questions* (San Francisco, CA, 2017)

Hampp, Constanze, and Stephan Schwan, 'The Role of Authentic Objects in Museums of the History of Science and Technology: Findings from a Visitor Study', *International Journal of Science Education*, Part B, V (2015), pp. 161–81

Janes, Robert R., and Richard Sandell, eds, *Museum Activism* (Abingdon, 2019)

Lindqvist, Svante, ed., *Museums of Modern Science* (Canton, MA, 2000)

Lubar, Steven, *Inside the Lost Museum: Curating, Past and Present* (Cambridge, MA, 2017)

Macdonald, Sharon, *Behind the Scenes at the Science Museum* (London, 2002)

McLaughlin, Hooley, and Judy Diamond, eds, *Science Museums in Transition: Unheard Voices* (Abingdon, 2020)

Morris, Peter J. T., ed., *Science for the Nation: Perspectives on the History of the Science Museum* (London, 2010)

Newell, Jennifer, Libby Robin and Kirsten Wehner, eds, *Curating the Future: Museums, Communities and Climate Change* (Abingdon, 2017)

Pantalony, David, 'What Remains: The Enduring Value of Museum Collections in the Digital Age', *HOST – Journal of History of Science and Technology*, XIV (2020), pp. 160–82

Pedretti, Erminia, and Ana Maria Navas Iannini, *Controversy in Science Museums: Re-Imagining Exhibition Spaces and Practice* (Abingdon, 2020)

Post, Robert C., *Who Owns America's Past? The Smithsonian and the Problem of History* (Baltimore, MD, 2013)

Rader, Karen A., and Victoria E. M. Cain, *Life on Display: Revolutionizing U.S. Museums of Science and Natural History in the Twentieth Century* (Chicago, IL, 2014)

Sandell, Richard, *Museums, Moralities and Human Rights* (Abingdon, 2017)

——, et al., eds, *Re-Presenting Disability: Activism and Agency in the Museum* (Abingdon, 2010)

Schorch, Philipp, and Conal McCarthy, eds, *Curatopia: Museums and the Future of Curatorship* (Manchester, 2019)

Science Museum Group Journal (2014–)

Thomas, Nicholas, *The Return of Curiosity: What Museums Are Good For in the 21st Century* (London, 2016)

Wood, Elizabeth, Rainey Tisdale and Trevor Jones, eds, *Active Collections* (New York, 2018)

Woodham, Anna, Alison Hess and Rhianedd Smith, eds, *Exploring Emotion, Care, and Enthusiasm in 'Unloved' Museum Collections* (Leeds, 2020)

ACKNOWLEDGEMENTS

For a small book, *Curious Devices and Mighty Machines* has incurred many debts. Work for Chapters Two and Three was supported by a Cain Short-Term Fellowship at the Science History Institute, Philadelphia. Some arguments have been rehearsed in Samuel J.M.M. Alberti, 'Why Collect Science?', *Journal of Conservation and Museum Studies*, xv (2017), pp. 1–10; and Samuel J.M M. Alberti, Elsa Cox, Tacye Phillipson and Alison Taubman, 'Collecting Contemporary Science, Technology and Medicine', *Museum Management and Curatorship*, xxxiii (2018), pp. 402–27; I thank the co-authors, peer reviewers and editors involved.

Although I present my own views and not those of my organization, National Museums Scotland has supported me in this endeavour, and in particular the Science & Technology department provided a convivial environment in which to work. Especially helpful among this team and elsewhere in the organization were Kate Bowell, Chris Breward, Rob Cawston, Adam Coulson, Julie Gibb, Sophie Goggins, Katarina Grant, James Inglis, Alan McIntosh, Elaine Macintyre, Xerxes Mazda, Molly Osborn, Ellie Swinbank, Tacye Phillipson and Louis Volkmer. So too the Centre for Environment, Heritage and Policy at the University of Stirling is a lively and supportive group that I am lucky to be involved with, and I have valued the collegiality of Chiara Bonacchi, Sally Foster, Siân Jones, Jennie Morgan and Holger Nehring.

Further afield, colleagues and friends in other organizations provided advice, information and endless details: at the Franklin Institute, Susannah Carroll and Jayatri Das; at Harvard University, Sara Frankel, Sara Schechner and Maureen Ton; at Norsk Teknisk Museum, Ageliki Lefkaditou and Henrik Treimo; at the Science History Institute, Hilary Kativa, Stephanie Lampkin and Erin McLeary; at the Science Museum Group, Alison Boyle, Robert Bud, Margaret Campbell, Paolo Cuocco, Selina Hurley, Jack Kirby, Nicole Simoes da Silva and Sophie Waring; at the University of Cambridge, Morgan Bell, Rosanna Evans and Joshua Nall; and at the University of Oxford, Janine Fox and Stephen Johnston. Elsewhere helpful colleagues included Anna Adamek, Ingenium; Kostas Arvanitis, University of Manchester; Richard Bellon, Michigan State

University; Caroline Blackwell, University of Newcastle, New South Wales; Malcolm Chapman, University of Glasgow; Simona Casonato, Museo Nazionale della Scienza e della Tecnologia Leonardo da Vinci; John Durant, MIT; Catherine Eagleton, University of St Andrews; Tom Everett, Ingenium; Sarah Gerrish, Sarah Gerrish Conservation; Scott Keir, University College London; Dmitry Kokorin, Polytechnic Museum; Catherine McCarthy, NISE Network; Sassy McGhee; Kathrin Ostermann; Katherine Ott, National Museum of American History; Ute Sampson; Klaus Staubermann, ICOM Germany; Rosemary Watt, Glasgow Museums; Marta Lourenço, University of Lisbon; Helmuth Trischler, Deutsches Museum; and several anonymous colleagues. Any errors, of course, are mine alone.

For helping not only in the aforementioned ways but also in reading drafts in detail and providing expert feedback I could not hope to fulfil, I am grateful to Stephen Allen; Rebekah Higgitt, National Museums Scotland; David Pantalony, Ingenium; and Alison Taubman, National Museums Scotland. Fiona Alberti's eagle eye improved every chapter and Tim Boon, Science Museum Group, read the whole thing and provided important feedback at the final stage. Finally, Margaret Clift read, critiqued, tolerated, cajoled and supported at every step of the way.

PHOTO ACKNOWLEDGEMENTS

The author and publishers wish to express their thanks to the below sources of illustrative material and/or permission to reproduce it:

Photo Samuel J.M.M. Alberti: 59; © Kristian Buus/Art Not Oil: 90; © 2015 Cognitive, www.wearecognitive.com: 86; © Collection of Historical Scientific Instruments, Harvard University, Cambridge, MA: 18, 19, 25, 56, 57, 68, 76; courtesy of Deutsches Museum, Munich: 13, 82 (photo Konrad Rainer), 87, 88; © Exploratorium, www.exploratorium.edu: 28, 29; Glasgow Science Centre: 6; The Henry Ford, Dearborn, MI: 85; History of Science Museum, University of Oxford: 22 (Inv. 13773), 24, 52 (Inv. 52173), 64; Gary Hodges for the NISE Network: 10; © The Hunterian, University of Glasgow 2018/photo Andrew Lee: 58; Ingenium – Canada's Museums of Science and Innovation: 14, 35, 41, 46, 54, 67; © Sergey Maymulov for Polytechnic Museum, Moscow: 3, 55; MIT Museum, Cambridge, MA/photo Samara Vise 2017: 7; © Musée des arts et métiers–Cnam, Paris/photo Philippe Hurlin: 17; © National Gallery, London: 21; photo courtesy of the National Museum of American History, Smithsonian Institution, Washington, DC: 92; © National Museums Scotland: 1, 2, 8, 11, 27, 31, 32, 37, 42, 43 (photo Dr Tacye Phillipson), 45, 47, 50, 51, 60, 61 (photo Julia Tauber), 62, 65, 66, 79, 81, 83, 84, 94, 98 (photo Neil Hanna); © National Science and Media Museum/Science and Society Picture Library: 91; © Norsk Teknisk Museum, Oslo: 73 (photo Lesley Leslie-Spinks), 74 (photo Geir Christiansen), 75 (photo Marie Skoie), 96 (photo Håkon Bergseth); Nuclear Decommissioning Authority/Dounreay Site Restoration Limited: 97; © Ontario Science Centre, Toronto/photo Tara Noelle: 5; Redorbital Photography/Alamy Stock Photo: 63; Russborough House, Blessington (on loan from the Apollo Foundation): 20; Science History Institute, Philadelphia, PA/photo Molly Sampson: 93; Science Museum/ Science & Society Picture Library: 9, 16, 23, 33, 38, 40, 44, 48, 49, 51, 53, 71, 72, 80, 89, 95; Science Museum of Minnesota for the NISE Network: 69; Smithsonian Institution Archives, Washington, DC: 15 (SIA_000095_ B42_F28_003), 26 (SIA_2010-2908), 30 (78-1202-33); Ken Stanek for the NISE Network: 70; © Stiftung Deutsches Technikmuseum Berlin (SDTB)/

INDEX

Illustration numbers are indicated by *italics*